Pneumatic & Hydrautic Conveying of Both Fly Ash & Bottom Ash

HOWARD G. LOWNES SR.

Brilliant Books Literary
137 Forest Park Lane Thomasville
North Carolina 27360 USA

TABLE OF CONTENTS

PREFACE

The author has endeavored to make this book useful to those who seek information regarding the kind of equipment and systems used to convey ash from power plants, the size of electric utility stations, or from an industrial plant of similar size. This book is a compilation of materials, equipment, design, and knowledge reflecting years of experience. It will discuss and describe some old concepts which are still in use today, as well as the more recent concepts.

It is not the intention of this book to make polished system design engineers out of all its readers, but to aid the engineer, or consultant with conceptual approaches. It is intended to present the various types of systems and equipment that make up an industrial system. It will also provide plant managers, operators and engineers ways to make improvements for existing systems.

I have spent over thirty years in engineering, design and research in the pneumatic and hydraulic conveying of boiler ash, working mostly with abrasive materials, and ash handling systems. Conveying abrasive, or nonabrasive materials all use the same principles, however, the materials being conveyed will determine the component parts based on the degree of abrasiveness of the material being conveyed.

The book is largely written around modern ash handling systems and equipment. There are several companies that have supplied systems for years, and we are seeing small companies entering the field. Each company has their own component parts that are standardized for their systems, but must be evaluated by the user.

The first nine chapters are devoted to hydraulic conveying system. Various types of systems and equipment are outlined and discussed. System design calculation and examples are presented to help understand system design, operation and power requirements. Pneumatic conveying is discussed in an additional eight chapters.

In writing this book I have attempted to provide a broad range of hydraulic and pneumatic conveying systems and equipment which I have personally designed as well as been involved with thought out my career.

CHAPTER 1

Introduction – Power Plant Ash Handling

Most ash conveyed from an electric utility station or from an industrial plant of any size is transported by a sluice, hydraulic, or pneumatic conveying system. Recently, drag chain conveyor systems have emerged for conveying bottom ash. By definition, a sluice or hydraulic conveying system is one in which the material is moved from one point to another enclosed in a sluice, or trough, or enclosed pipe by a liquid, and in most cases this is water. A pneumatic conveying system is one in which the material is conveyed by a gas, which in most cases is air, inside an enclosed pipeline.

Figure 1 shows the various areas of a power plant where ash is deposited, thereby requiring a form of materials handling system. Ash, as we know it, is defined as the residue from the burning of coal, lignite, wood or peat. Burning of oil in large boilers, also produce some ash, but this will not be addressed in this book.

Pyrites, or mill rejects have been shown in the figure as ash, primarily because they are conveyed by the ash handling system, and are very abrasive.

The figure shows the ash produced by a pulverized fuel boiler (P.C.), and does not hold strictly true for all types of boilers. It does show that the heavy ash and larger particles are deposited first, followed by ash in the descending order of size with the stack being the finest.

The first part of this book will be devoted to hydraulic conveying of ash. We will first look at some old methods, which could be still in service today, then the more resent methods used over the past few decades. The second part of the book will be devoted to pneumatic conveying of boiler ash, commonly known as fly ash.

This book is intended to show system concepts and components, but not intended to make experts for system design. Each plant, location, and system is unique in it's self.

Also a few system design calculations and examples are presented to aid design engineers to understand system operation and power requirements.

This book is an attempt to show a broad range of hydraulic and pneumatic conveying systems which I have been involved with personally for over thirty some years.

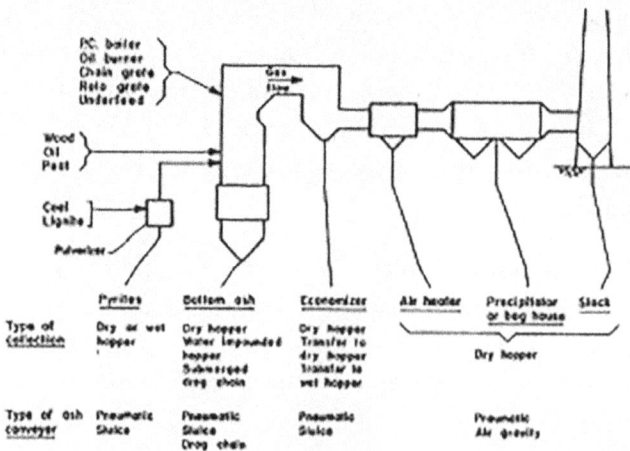

Figure 1

CHAPTER 2

Sluice Systems – Hydraulic – Sluiceways

In the early systems, the common application was to convey the heavier ash, or bottom ash by the action of a high velocity stream through a sluice system where the ashes are introduced. The ashes are flushed from the ash pit by a hand-controlled jet, and the mixture flows into the sluice, along the floor of which a high-velocity impelling jet is directed. Figure 2 shows an old style ash system with a gate holding the ash in the ash hopper, and a nozzle in place to aid in the removal of ash from the hopper. Figure 3 shows the ash being removed and flowing down into the sluice system. These sluiceways usually are just below floor level and are stepped at intervals to accommodate booster jets. The sluiceway is lead to a sump pit. Some of these pits where located outside of the building for the convenience of using a grab bucket to transfer the ashes to either a truck or rail car. Others were located within the plant walls, and the ashes were either pumped directly to an ash disposal pond located away from the plant, or to an overhead bin located outside of the building. In the case of the overhead bin the water was drained off then the ashes were dumped into either a truck or rail car for further disposal.

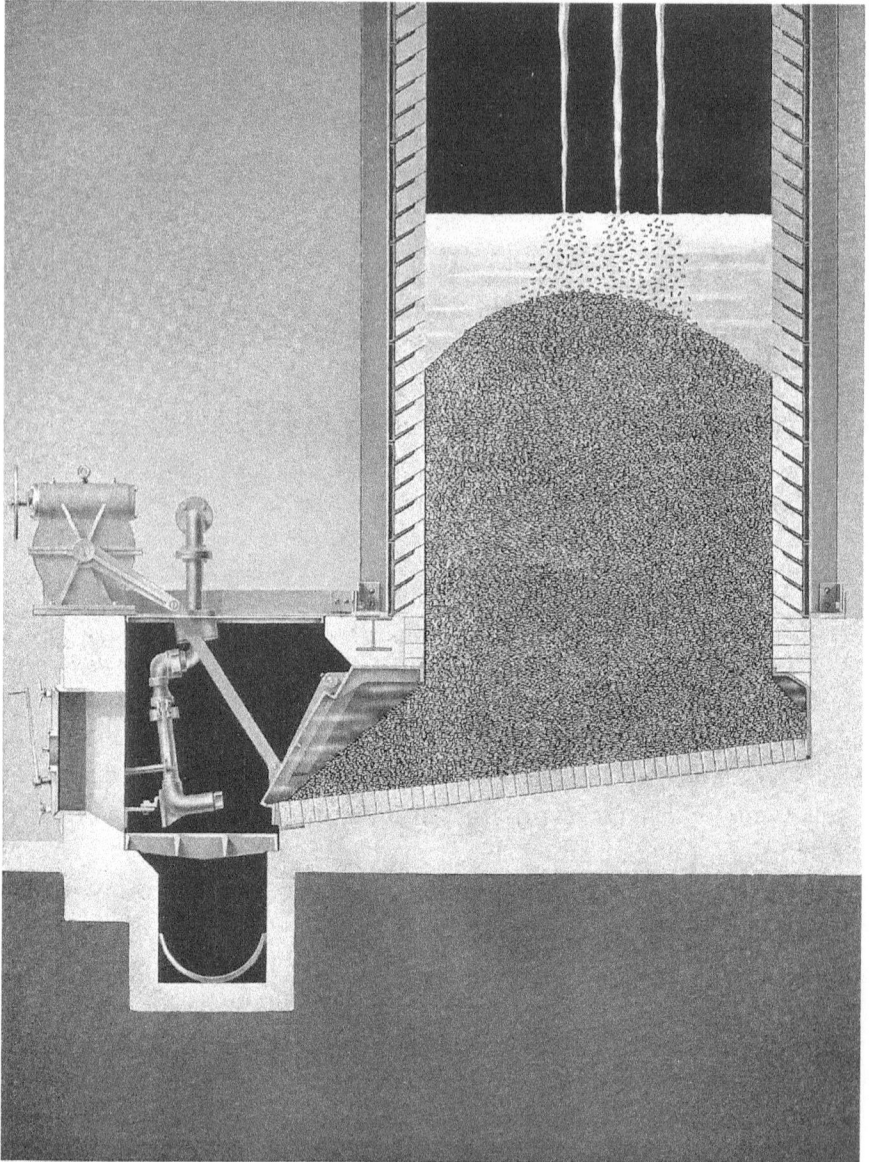

General Arrangement of Hydrojet System for Continuous Slagging Furnaces.

Figure 2

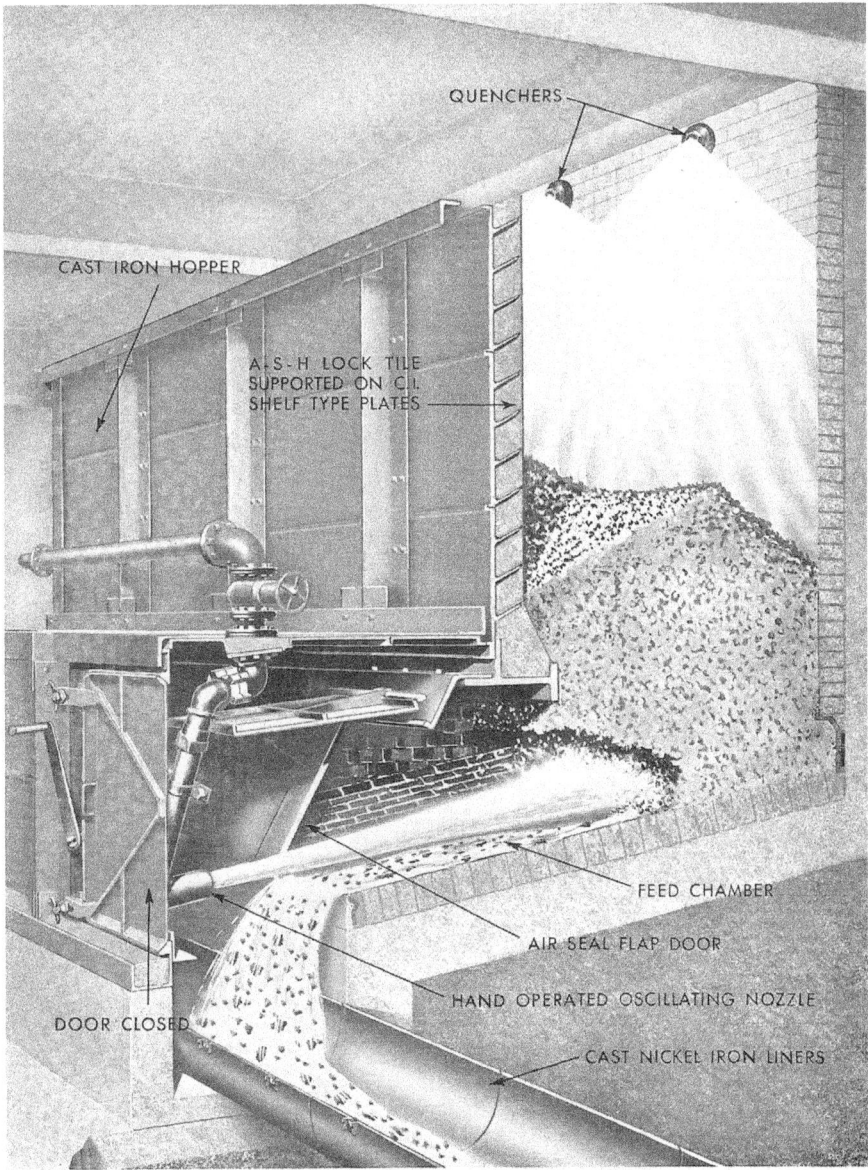

QUENCHERS

CAST IRON HOPPER

A.S.H LOCK TILE
SUPPORTED ON C.I.
SHELF TYPE PLATES

FEED CHAMBER

AIR SEAL FLAP DOOR

HAND OPERATED OSCILLATING NOZZLE

DOOR CLOSED

CAST NICKEL IRON LINERS

General Arrangement of Hydrojet System for Chain Grate Stoker Fired Furnaces.

Figure 3

These systems had an open sluice on a slight downward slope, or stepped, as mentioned above. High pressure water flows through a series of nozzles along this sluiceway. The sluiceway consists of a horizontal concrete trench extending beneath the ash basement floor. The sides and bottom of this trench are protected against wear by a series of half round, nickel cast iron liners, placed end to end. The method of installing and securing these liners is unique in that no nuts or wedges are used. The ends of two turnbuckles are dropped into lugs, which are integrally cast on the liners. The turnbuckles are then tightened and since the liners are somewhat thinner at the edges than they are at the bottom, the liners contract. They are then placed in the bottom of the concrete trench so that the mitered edge of the liner overlaps the mitered edge of the adjacent liner, as shown in Figure 4, and then the turnbuckles are loosened and removed. The liners thus expand solidly against the side walls of the trench and cannot be moved in operation, but they can be easily removed by using the turnbuckles as a tool, as previously described. The floor of the sluice is stepped at intervals to accommodate booster jets as shown in Figure 5.

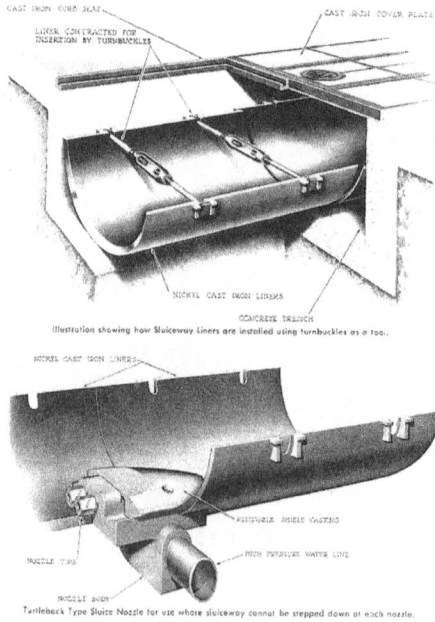

Figure 4

Ashes are removed at the rate of 1 to 2 tons per minute and since they are fed into the discharge sluiceway in proportion to its carrying capacity, it is impossible to overload the system. Depending on conditions, about 3 to 6 pounds of water are required per pound of ash removed and since this water in the average installation, is supplied with high pressure water was usually at 100 PSI and produced a velocity in the sluiceway of about 130 feet per second. This high pressure water provided the motive force to move the ashes and also the advantage that it reduces the wear on the liners since the velocity causes the ash to ride on the water, instead of rolling along the bottom. Ash laden water from one trench, strikes a renewable abrasion resisting wear plate and dropping into the next trench. See Figure 5. This eliminates the wear on trench liners, which would naturally occur if they were laid in a curve.

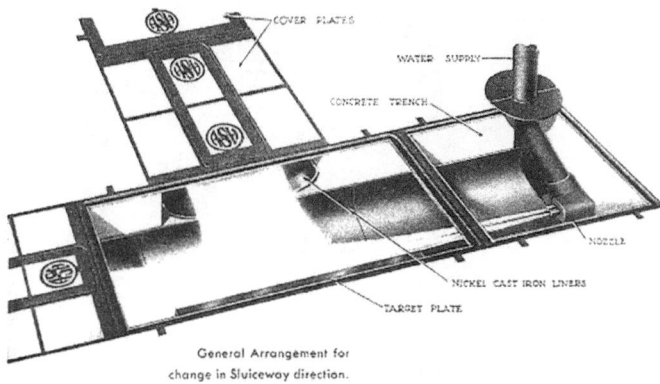

General Arrangement for change in Sluiceway direction.

Figure 5

The entire length of the sluiceway is enclosed, level with the floor, by means of removable cover plates, which rest in cast iron curb seats, also shown in Figure 3.

Discharge nozzles are located at intervals along the sluiceway as required to maintain the velocity of the ashes and are always located back of the point where a sluiceway changes direction, or joins another sluiceway. Depending on the design of the sluiceway, if liners could provide a 3" step, the nozzles are located such that they are protected from wear. If steps are not an option, then the nozzles are in the bottom of the sluice liners and are protected by a replaceable wear section above them. In Figures 6 and 7, shows the sluice liners with the various types' of nozzle arrangements. Booster nozzles are located approximately every 75 feet along the sluiceway.

There are some of these systems still in use today.

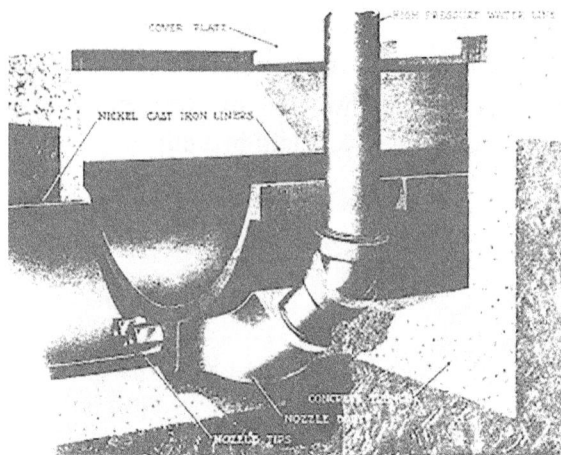

Figure 6

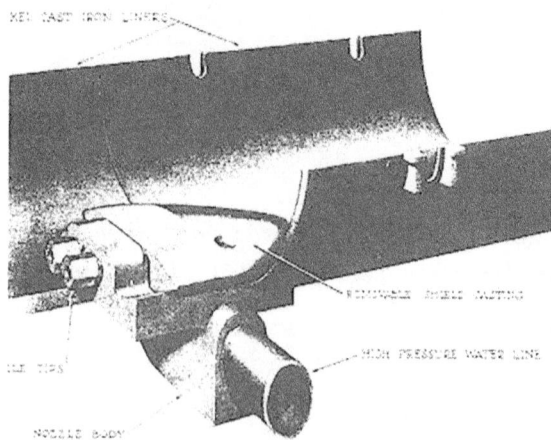

Figure 7

CHAPTER 3

Sluice Systems – Hydraulic – Closed Pipeline

The current systems for sluice systems is through an enclosed pipeline, and can be defined as the movement of heavy solid materials, such as bottom ash and pyrites, or mill rejects. This book will consider the movement of material in a totally enclosed pipeline.

Bottom ash from all but extremely small boilers, under 250,000 to 500,000 pounds of steam per hour, has traditionally been collected in water impounded hoppers, which will be discussed in another chapter. Those boilers in the rage of 250,000 to 500,000 pounds of steam per hour have been conveyed by negative pressure conveying systems, which also will be discussed in another chapter. See chapter 11. This ash is fed into an adaptor or sump which flows into a pumping device. The pump may be a centrifugal materials handling pump, or it may be an ejector (jet pump). Figure 8 illustrates a centrifugal pump. Figure 9 illustrates a typical jet pump.

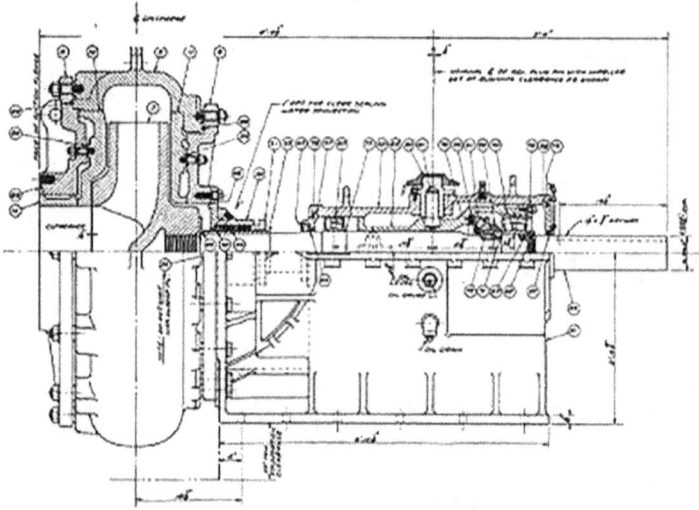

Figure 8

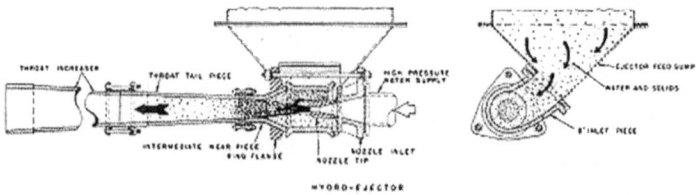

Figure 9

In applying either pump, considerable dilution is required at the suction in order to provide a slurry, that can be pumped. Usually the solids represent about 20% of the total weight of the slurry pumped. The suction of a jet pump is limited by the system head which also determines the amount and pressure of the jet impulse water.

This chapter will discuss the methods of conveying this material to the disposal area. There are these three methods that are currently used for pumping this material.

    a) A single jet pump located at the discharge of each outlet.

b) An inline material handling pump, taking its suction from each outlet.

c) A combination jet pump-material handling pump system.

The jet pump has the ability to be self-regulating. That is, when the discharge pressure equals or exceeds the suction pressure, the jet pump will not accept any more material; or water. This feature alone tends to regulate the amount of material in the sluice line. Some people think jet pumps can operate with a closed off head or, alternately, accept water or air without affecting it. See Figure 10. Jet pumps must be checked for capitation just like centrifugal pumps. Its one bad feature is that it requires high pressure water for operation, 100 upwards to 350 PSIG, and is only about thirty percent efficient. Both the materials handling pump and the jet pump require added controls to insure a flooded suction at all times. The centrifugal pump is becoming more popular, but at times can restrict the operator's action in regards to lowering the hopper water level. See Figure 11.

Jet pumps have a limited head they can develop. When pumping long distance, the use of a jet pump discharging to a transfer tank and a centrifugal pump discharging from the tank allows the best of both systems. First, the jet pump only requires about 100 PSIG pressure, but will allow the operator complete freedom of lowering the ash hopper water level with less concern if the suction water to the jet pump was reduced to below suction requirements. Second, the centrifugal pump only requires a simple level control at the transfer tank. See Figure 12. This system is not as efficient as other approaches.

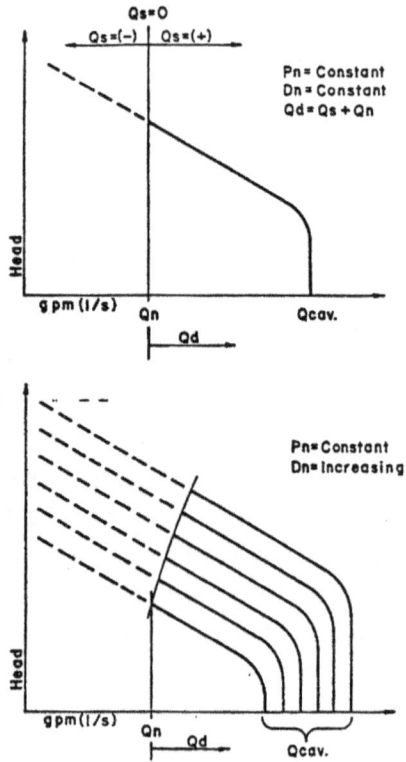

Figure 10

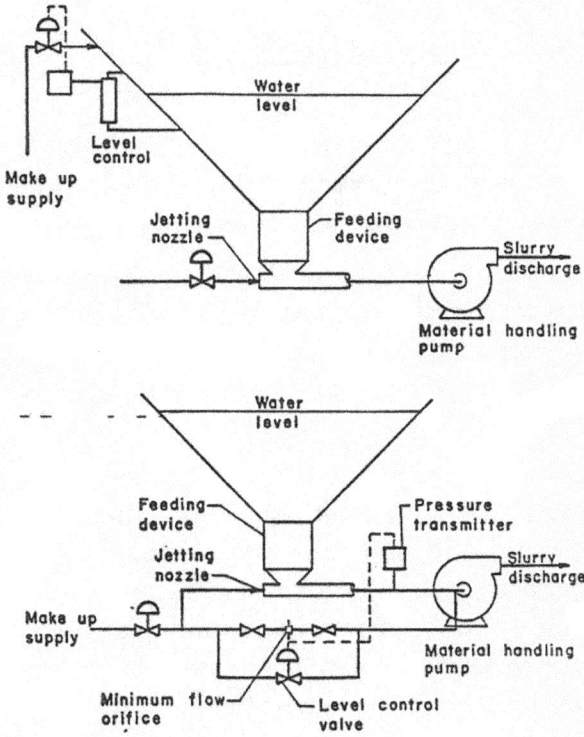

Figure 11

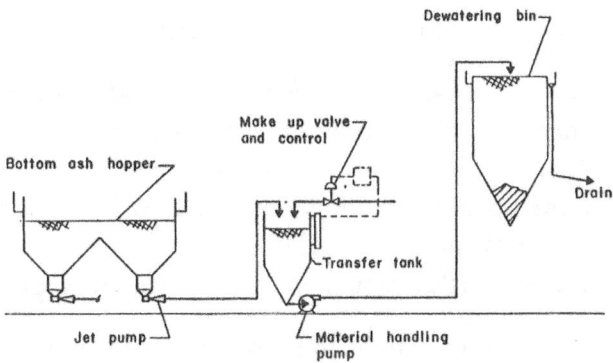

Figure 12

The terminal point of an ash conveying system will be either:

1. A low area where the ash can be deposited, and dried out by permitting run-off of the conveying water.

2. A pond, or lagoon, in which the ash is allowed to settle to the bottom, with excess water eventually over-flowing to adjacent natural streams. This is discussed later.

3. A dewatering bin into which the ash-water slurry is pumped. Ashes are settled and the water is removed through decanting and dewatering elements, so that relatively dry ash is delivered to trucks or railroad cars. Dewatering bins will be discussed in Chapter 7.

Where ground conditions permit, ashes can be pumped into natural or artificial ponds. If the ground is impervious, man-made ponds or lagoons can be constructed in such a manner that water can be recovered, stored and returned to the plant for reuse in the ash conveying systems. To be effective these storage facilities must cover a considerable area since retention time is the only means by which ash can be settled and separated from the conveying water. Where space permits, volume in a storage basin should be provided for one day's retention of the ash conveying water. Fly ash should not be pumped into a pond from which water is to be reclaimed because of its extremely slow setting rate, unless the pond is greatly enlarged to provide long retention time.

If water is not to be reclaimed, ponds can be considerably smaller than otherwise. Where ponds or pits are used for temporary storage of ash, they should be divided into two compartments so that one side may be cleaned while the other is receiving ash. Overflow from such ponds or pits will flow to sewers or to clarifiers depending on regulations. In most localities untreated overflow is no longer permitted. See Figure 13

For further discussion of material disposal in natural or artificial ponds refer to chapter 18.

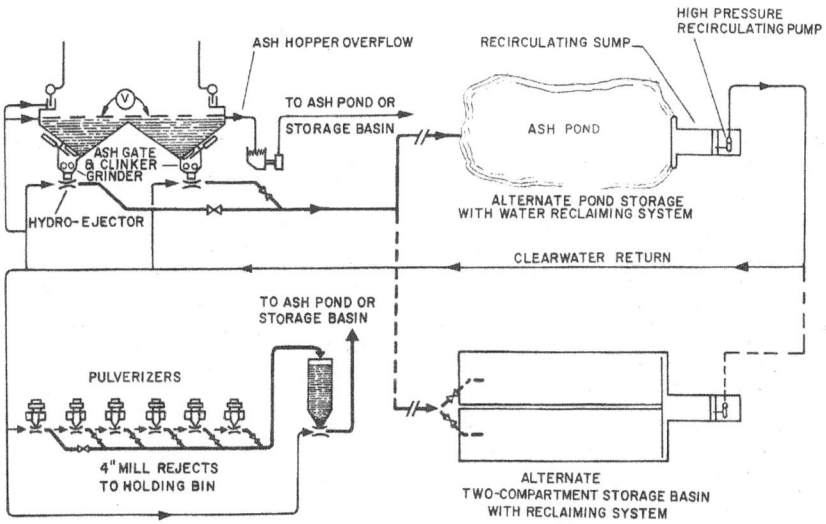

Figure 13

CHAPTER 4

Mill Rejects Systems

There are many different types of Mill Rejects Equipment, depending on the supplier of the boilers that are in service, or being built. They all reject pyrites, or items they can't crush, or pulverize. Storage hoppers are mounted below each mill discharge spout to remove the rejects continuously. This material is most conveniently handled hydraulically as an adjunct to the bottom ash system. High level indicators initiate the removal operation with rejects removed in sequence from each hopper to a central holding bin by means of suitable sized jet pumps. Discharge may be through individual pipes or through a common pipe as plant conditions dictate. Storage hoppers should have as large capacity as possible to reduce the frequency of operation. Hoppers are normally constructed of carbon steel with overflow seals designed for the maximum possible mill pressure. Special corrosion resistant metals can be used when specified.

Mill discharge control valves or gates are sometimes supplied by the pulverizer manufacturer. When this is not the case, cylinder operated gates are supplied as part of the mill reject system, but specifications should be clear as to the extent of this requirement and the degree of interlocking control to be included as part of the ash handling system.

The holding bin into which rejects are discharged should have sufficient capacity so that it can be emptied on the same cycles as the bottom ash hopper serving the same boiler unit. See Figure 13, in Chapter3. A jet pump of the same size as those at the bottom ash hopper is used to empty the holding tank and to discharge mill rejects into the same line through which bottom ash is conveyed. Because of the heavier nature of mill rejects, a substantially greater flow velocity may be required when conveying rejects as compared to the flow velocity when handling bottom ash.

Where the holding tank location is close to an ash hopper using a centrifugal pump, it is possible to extend the pump suction line, by separate valves, to the holding tank. In such a case, it is necessary to pump bottom ash at the higher velocity required by the mill rejects since pump impeller speeds are not conveniently change to mach varying flow requirements.

Discharge of mill rejects into the bottom ash hopper is no longer considered practical. Damage to boiler tubes caused by splashing from the mill rejects discharge pipes and the lack of uniformity in density of the resultant slurry which causes line plugging problems has eliminated this method of handling of mill rejects in most cases.

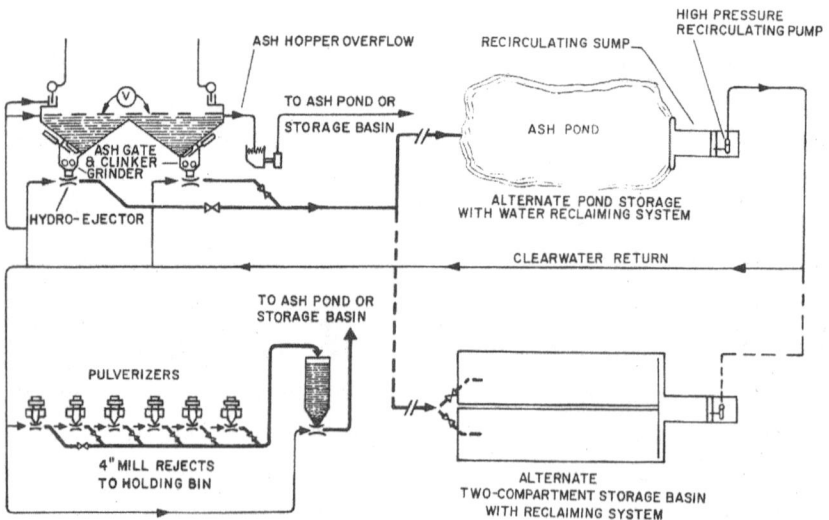

Figure 13 (repeated for reference)

CHAPTER 5

Bottom Ash Hoppers

Bottom ash from all but extremely small boilers, under 250,000 to 500,000 pounds of steam per hour, has traditionally been collected in water impounded hoppers and transported to a disposal facility by a sluice system. Those systems under 250,000 to 500,000 pounds of steam per hour have been conveyed by negative pressure conveying systems. See chapter 11.

The size of the modern boilers requires hoppers with large storage capacities, steep, self-feeding slopes, and water containment to quench ash and to provide delivery at a high rate to the conveying system. Figure 14 shows a typical ash hopper for a large, dry bottom utility boiler, which is rectangular in shape, is designed and sized to shatter, collect, cool and to store ash for twelve hours or more. Consideration must be given in the design of these hoppers to provide adequate slopes for gravity feeding of any masses of ash or slag that may fall from the boiler. The hoppers sole connection to the boiler is through a seal plate water trough arrangement. The trough is located around the entire periphery of the hopper and allows expansion and movement to occur, as well as seals the boiler gases against escape. See Figure 14. Most of the larger hoppers are shaped in the form of a "V", or multiple "V's", depending on the overall boiler size. A three "V" hopper is used when the boiler has a center dividing wall to allow the added material shed from the wall to fall into a deeper water area. The conventional design shown in Figure 14 may be modified, to provide for unusual conditions of slagging or exceptionally high ash content, by increasing the hopper width under the seal troughs to give greater storage capacity, or by increasing the length sufficiently to bring the end outlet openings closer to a line under the furnace end walls from which the largest slag masses can be expected to fall. Discharge openings, placed on the bottom slopes, aid in the unrestricted flow of ash from the hopper. Adequate clearance around all parts of an ash hopper is essential for maintenance purposes.

Each of the "V" sections or discharge pant legs contains at least one discharge door inside a water tight enclosure. The enclosure supports a clinker grinder which connects directly to the sluice system piping. The entire inside surface of the hopper is refractory lined which absorbs the impact loads and abrasion from the falling and moving ash. The refractory is water cooled above the water line to prevent spalling. To aid in the discharge of material from each pant leg, a series of flushing nozzles located across the top of the sloping legs are used to direct the material towards the discharge door.

There are two accepted methods of discharging the ash from this type hopper. One method is to lower the water level each time the ash is discharged, and the other method is to maintain the water level. If the slopes of the pant legs in this type hopper are steep enough, then you normally can maintain the water level. Maintaining the water level minimizes the exposure of the refractory to impact forces of falling ash and the direct radiation from the boiler.

The space between the boiler and the hopper is closed by a water seal trough and seal plate. The seal plate is part of the boiler structure. The water seal permits expansion of the boiler without placing stress on the ash hopper. A series of weir boxes or similar devices around the top periphery of the hopper permits the overflow of a continuous curtain of cooling water over the ash hopper refractory lining thereby keeping the refractory wet and avoiding the possibility of spalling at the water line.

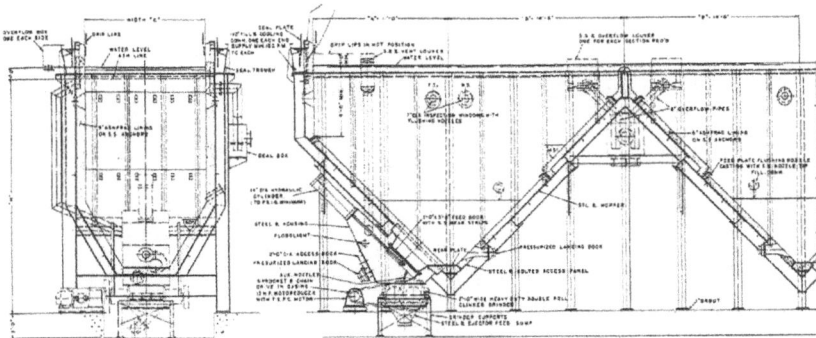

Figure 14

The refractory cooling elements and the parts of the water seal troughs exposed to furnace gases and high temperatures are constructed of stainless steel or other corrosion and temperature resistant metals.

As a rule, thermal shock on an ash hopper will be minimized if the width inside the refractory lining is at least three times the boiler throat opening. With refractory cooling it is generally possible to maintain safe wall temperatures with a monolithic wall lining 9" thick. Bottom slopes are adequately protected with 6" lining. Firebrick or other types of block refractory are not recommended because they have a tendency to lift out under temperature changes and cause obstruction to the flow of material from the hopper.

Poke doors, observation windows, access doors and panels are available to suit all specified requirements.

For smaller boilers with a steam generating capacity in the 1,000,000 lb. per hour range or smaller, where automatic feeding of ash is not a consideration, flat-bottom ash hoppers can be supplied with a single outlet at one end or at the center. Figure 15, illustrates the hopper with the discharge at the end, and Figure 16 illustrates a hopper with the discharge at the center which provide maximum storage within limited headroom. Most ash must be fed from the flat bottoms by means of nozzles requiring operator attention. Such ash hoppers are limited in length to about 35 feet.

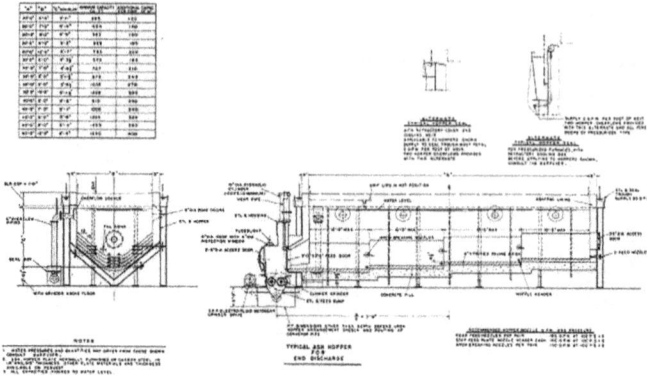

Figure 15

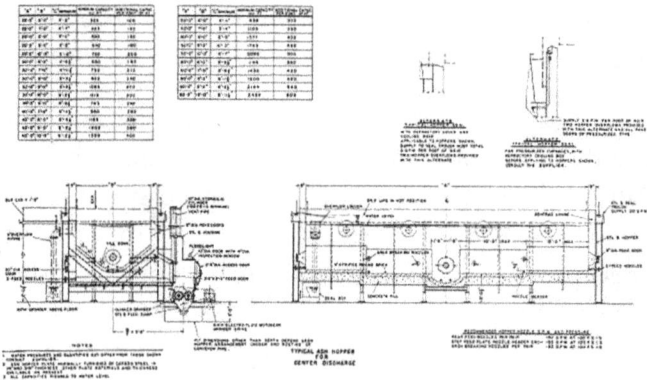

Figure 16

Slag Tanks are for continuously slagging wet –bottom boilers. The slag tanks are supplied as illustrated in Figure 17, when slag enters the tank through a throat section which is hung from the boiler. Near the top of the throat is a water-cooled arm which swings back and fourth periodically to cut off any strings of chilled slag which may accumulate near the point of discharge. The slag then drops into a pool of water where agitating jets assist in the granulation process. A trough around the upper periphery of the throat admits a curtain of water down the sides of the throat lining to prevent slag from sticking to it. A water seal trough is provided between the throat and the slag tank to prevent movement of the throat due to expansion. Feeding of slag from the tank into a clinker grinder and pump suction connection is similar to that described for a dry bottom ash hopper. Refractory linings are generally of the thickness shown in Figure 17.

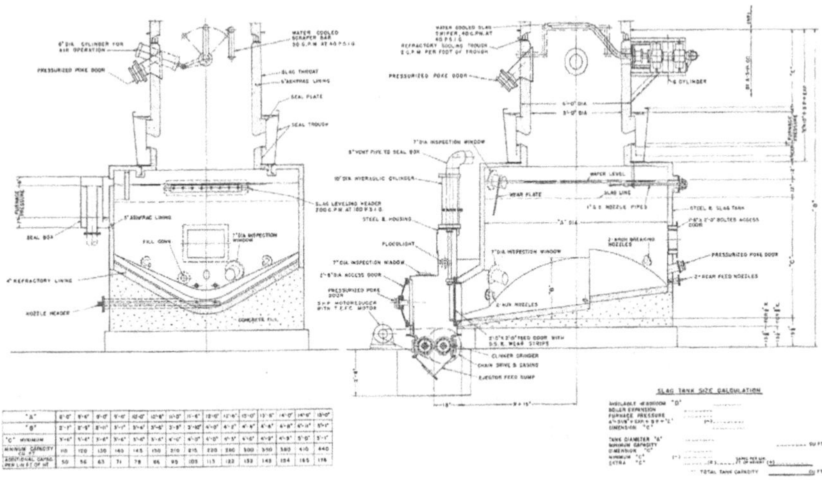

Figure 17

Discharge gates for any large boiler should include discharge openings large enough to pass the largest piece of ash or slag that may fall into the hopper from the furnace. Figure 18 shows the arrangement of a feed gate available for water-filled ash hoppers. The 3' × 3' gate is recommended for all large utility boiler ash hoppers. There is no lower limit of boiler capacity to which this gate might not be applied but in lower capacity ranges gates are as small as 2' × 2' are available. Water-tight carbon steel enclosures are normally supplied around discharge gates, equipped with access doors, observation ports and flood lights. If corrosive conditions exist special metals can be used for gate enclosures. Air-water or air-oil converters are available to supply fluid-power to the discharge gate operating cylinders. Plant air at 80 to 100 psig is required.

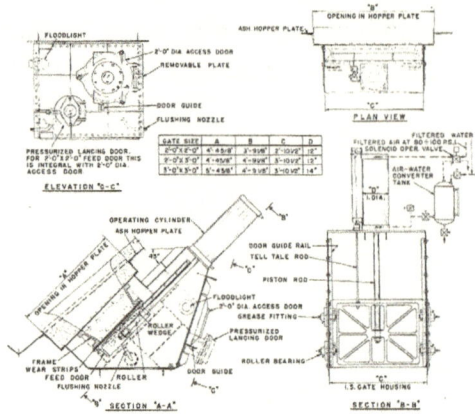

Figure 18

Clinker Grinders, are double roll units and are provided in two principal sizes to meet various conditions of capacity and masses of slag or ash that must be crushed. A clinker grinder must be selected of a size capable of crushing and feeding ash at a guaranteed rate without extreme fluctuations on either side of the average rate. Figure 19 shows a high capacity double-roll clinker grinder with a capacity up to 150 tons per hour.

Figure 19

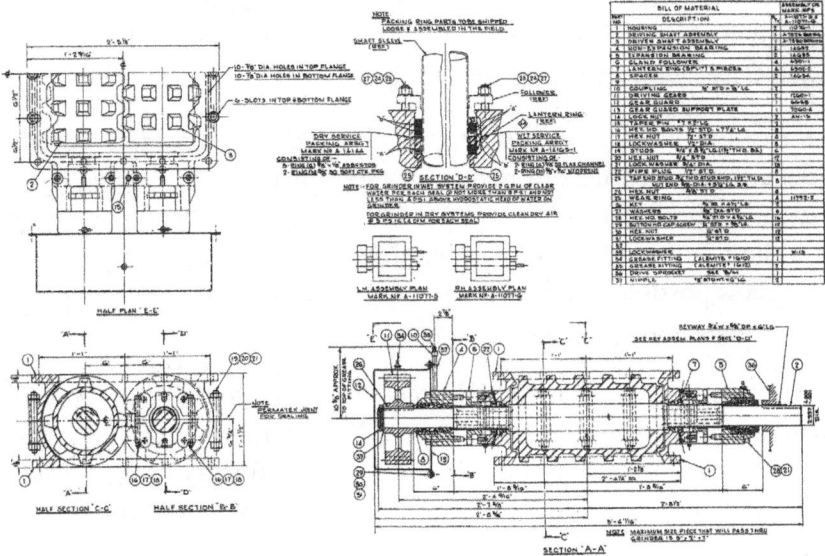

Figure 20

Figure 20 shows a smaller unit used where capacity requirements are 75 tons per hour or less. Both machines are for heavy duty ash and slag crushing. Drive sizes vary from 5 HP to 25 HP depending on material to be crushed and required system capacity. The high capacity grinder is always required where 3 ft. wide hopper discharge gates are used. The smaller grinder is generally applied to slag tanks and to ash hoppers under moderate sized steam generators where 2' × 2' discharge gates are considered adequate. With crusher rolls on fixed centers all ash is positively broken into particles about 2" in diameter, a size which will pass through ejectors, or pumps, and the conveying line without plugging. Specifications should always require fixed passages through clinker grinders to prevent the discharge of particles too large to pass through the conveying system. Crusher teeth are cast on the rolls and when worn, may be built up to original size by welding rods. Work hardening manganese steel crusher rolls and teeth provide the most durable material for ash or slag crushing.

The use of the largest available grinders and ash hopper discharge gates is good insurance against costly outages which may occur with the variation in quality of today's fuels. This is particularly true in the case of large boilers where the characteristics of the ash cannot be predicted in the design stages with any degree of certainty.

In the case of gravity feed ash hoppers on large units it may be desirable, also as an insurance measure to provide a second discharge gate and grinder at each hopper cavity. This will permit adequate time to maintain one set while the second system is operating. In addition, this arrangement also makes possible the installation of a standby ash discharge line with its obvious advantages.

Where light ash loads are anticipated single roll crushers are available. These are essentially the same as the 2'–0" wide double roll grinder except that one crusher roll is replaced with a stationary breaker plate. A single roll crusher should be used only when the system handling rate is to be 20 tons per hour or less.

CHAPTER 6

Chain Drag Conveyors

Traditionally, in this country, bottom ash has been transported from the ash hopper either hydraulically or pneumatically, and it has only been in the recent years that interest has grown in mechanical systems, primarily submerged drag chain conveyors, have been the preferred method of bottom ash conveying.

Mechanical Drag Conveyors typically consists of a submerged drag chain conveyor located directly under the boiler. The ash falls directly into a water impounded trough, is quenched, and is then pulled up an incline by the conveyor flights allowing it to dewater. The conveyor discharges into another conveyor or into portable containers for removal to an intermediate or permanent disposal site. The advantages of this type of system include the following: lower kilowatt requirements, substantially smaller volumes of water compared to a sluice system, minimum amount of on-site real estate, immediate quenching and dust suppression of the ash on a continuous basis, and a variable rate of removal, if desired. In addition, the continuous movement of the ash, by the flights, eliminates the possibility of hardening due to the calcium oxide content.

The mechanical drag conveyor, see Figure 21, has varying rates of removal. The horizontal section of the conveyor, located directly under the boiler, is as long as the bottom of the boiler, and as wide as required, with a 39 degree inclined dewatering section. The ash discharges from the top of the inclined section into an enclosed chute leading to a belt conveyor. The trough type belt conveyor carries the material to a disposal site, or discharged to the top of a silo. The silo is a conical live bottom design to facilitate flow of the damp bottom ash.

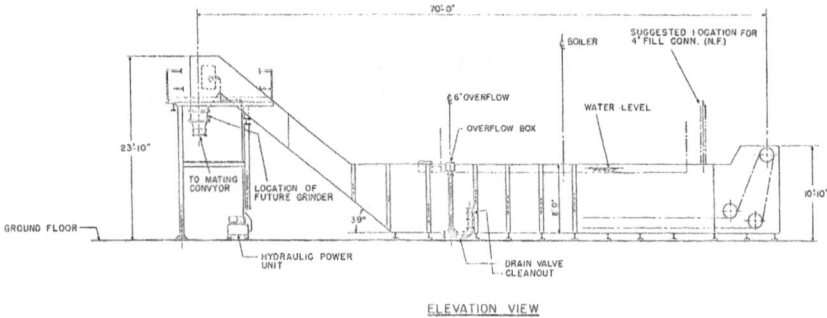

ELEVATION VIEW

Figure 21

The main component of the conveyor is the chain itself. This is a bar-loop type chain, having induction hardened wear surfaces, i.e. the connecting pin, inside radius of the loop link and the top and bottom edges of the side bars. Removal or changing of links is accomplished easily by removing several steel pins, similar to cotter pines, and changing the required components. A flight attachment designed specifically for this chain is bolted to each strand of chain and the flight is then bolted to the attachment. The flight is a fabricated, reinforced, structural member with abrasion resistant materials welded to the top and bottom surfaces. When the wear materials have deteriorated to the extent that replacement is necessary, it can be accomplished by burning off the remaining wear material and re-welding new wear strips. It is not necessary to replace the entire flight.

The upper, conveying trough is water tight with the bottom surfaces (dragging surface) lined with easily replaceable wear plates. The chain return trough, underneath the conveying trough, is basically dry and has two wear tracks which support the flights for the entire return path. These tracks are also easily replaceable.

The chain is driven by two hardened steel sprockets mounted on the alloy steel drive shaft. Heavy duty roller bearing pillow blocks are used to support all the through shafts in the conveyor. Specially mounted bearings with water flushed seals are used to support the four submerged idlers. The take-up is a manually adjusted screw type take-up mounted in the horizontal position at the back end of the conveyor. The conveyor is driven by an hydraulic power unit. The hydraulic pump is driven by a 10 HP electric motor, mounted directly to the high speed side of shaft mounted gear reducer. This drive arrangement is a very compact package which eliminates the need for massive drive supports as required by a foot mounted gear box with a chain and sprocket drive. In, addition, the hydraulic package provides over-torque protection for the conveyor by allowing the fluid to by-pass the motor through a relief valve in the event the conveyor stalls.

Controls for the conveyor are very simple, see Figure 22. In addition to the normal start and stop, a speed switch is provided on the take-up shaft to sense a lack of chain motion. In the event of a stall, this switch will automatically stop the conveyor drive and alarm the operator. Water temperature controls, water level controls, and chain tension indication are also provided. A series of steam spray nozzles activated intermittently by a limit switch sensing the presence of a flight; have been included to minimize the build-up or carryover of material by the flight to the return trough. However, this is not 100% effective and a certain amount of ash does transfer to the lower rough and deposit in the back end, creating a housekeeping problem.

The boiler seal is provided by a rectangular skirt extending from the bottom of the dry ash hopper into the water filled conveyor trough.

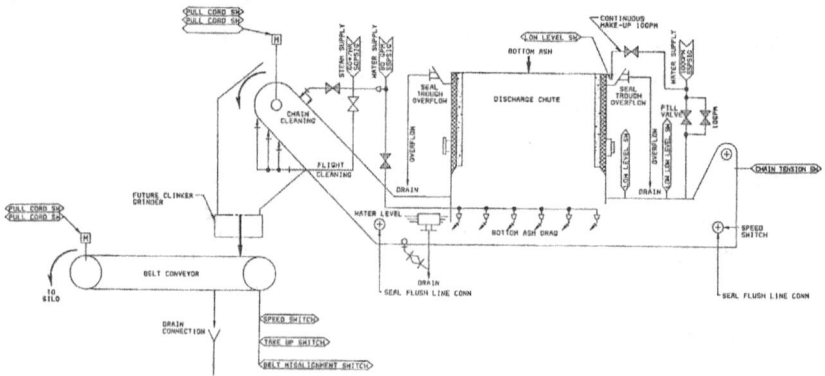

Figure 22

CHAPTER 7

Dewatering Bins

Dewatering bins are essential to closed-loop hydraulic ash handing systems where environmental regulations prohibit the discharge of ash-contaminated water into any bodies of natural water. In such cases, dewatering bins become an integral part of recirculating systems which will be described later.

Figure 23 shows one group of standard dewatering bins. Standard bins are also available in a variety of capacities to meet the storage requirements of plants of any size. Capacities range from 25 tons to 1,000 tons of ash having a dry bulk weight of 45 lbs. per cubic foot.

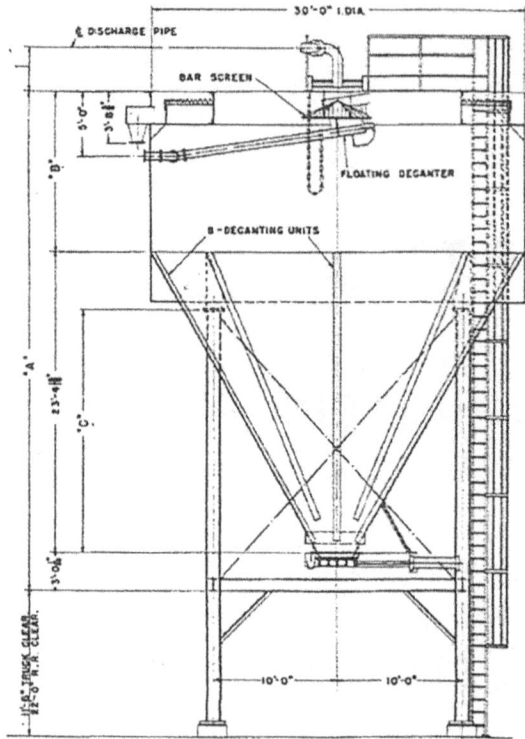

Figure 23

For most effective operation, two dewatering bins are normally required, one available for receiving ash, while the other is being dewatered and unloaded. Ash is pumped into a bin over a bar screen, which permits the finer material to drop directly into the center, while the coarser particles are diverted toward the sides of the bin to form a filter to trap fines before they can reach the decanting elements. A center decanting element is also available for decanting the center section of the bin.

When the bin is filled to the top, excess water overflows a serrated weir, around the bin periphery, into a trough from which it flows by gravity through a drain pipe to waste or to a settling system for recycling. An important element in any dewatering bin is an underflow baffle, concentric with the outer shell. All material entering the bin must pass under this baffle before reaching the overflow weirs. The baffle also prevents the turbulence caused by incoming conveying system discharge from reaching the weirs so that flow over the weirs is as steady and undisturbed as possible. Dewatering bins are constructed of carbon steel plate of thickness depending upon bin diameter and loading conditions. The lower cone section forms an angle of 60 degrees with the horizontal. Provision can be made, on the cone for installation of vibrators where ash is expected to be sticky or otherwise difficult to flow. Special metals can be utilized if specified, to combat unusual corrosive conditions.

Ash may be pumped into a bin in several cycles until its maximum storage level has been reached. This is usually determined from experience, but level detectors are available to indicate full-load conditions by means of a light on a remotely located control panel, if desired.

After filling a bin to its rated capacity, flow must be diverted to a second bin where the filling procedure is continued. The first bin is allowed to stand undisturbed until ash has settled leaving relatively clear water above the pile of ash, usually about one hour. This "top water" is then drained off by means of a floating decanter which rides down on the surface of the water until solid ash is struck. At this point the valves controlling the stationary decanting elements on the bin are opened to permit controlled, slow draining of water from the body of ash in the bin. The stationary decanters on the sides and the center if installed, of a dewatering bin are faced with stainless steel screens of the self-cleaning type with openings usually about 0.06'. By controlled low velocity discharge most fines are trapped within the body of ash in the bin. The total dewatering process time will vary depending on the quality of the ash and the degree of dewatering desired. Normally, about eight hours are sufficient to obtain a product satisfactory for removal in trucks or railroad cars. Ash is discharged through the bottom by means of a hydraulically operated gate with a nominal opening of 3'-0" in diameter. This gate is provided with an inflatable sealing tube to prevent leakage between the frame and the gate. In the event any leakage does occur, the gate is so designed that water is directed to a drain trough thus assuring that no water can be spilled to the ground. Gates are equipped with thermostatically-controlled electric strip heaters to prevent freezing if climatic conditions indicate this possibility.

Dewatering bin sizing, in addition to providing for decanting cycles, generally is determined by the longest period during which it may not be possible to unload them. This is usually considered to be a weekend of sixty-four hours. Where ash is known or is anticipated, to have cementing properties, it may be well to consider the use of a number of small bins and a procedure of frequent unloading to prevent ash from setting up with consequent removal difficulties.

Controls are provided to regulate the decanting rate to achieve the degree of dewatering and the least amount of carryover for any type of ash. Valves and operators are an integral part of the dewatering control system, but drain and overflow pipes are not normally included.

CHAPTER 8

Settling and Surge Tanks

When it is necessary to conserve water, or where regulations prohibit the discharge of any ash contaminated water into sewers or streams, fully closed loop recirculating systems are added to the hydraulic ash handling systems to permit reuse of a large percentage of the conveying water. Overflow from ash hoppers and dewatering bins is caught in a settling and storage tank system. Water is first allowed to enter a shallow settling tank usually of a diameter substantially greater than the dewatering bins, as the water flow enters the settling bin it goes under a baffle prior to going over the weir then to the storage tank. This permits particles to drop out prior to going over the weir. Since the settling tank is greater in diameter it allows for a greater weir length so that overflow is further reduced and fine particles which escaped from the dewatering bins can drop out and be recovered as sludge. The water finally is drained to a storage tank for reuse in filling ash hoppers and dewatering bins and for use in conveying of ash.

Water is returned to the ash conveying system by centrifugal pumps. Pumps for intermittent service are usually for higher pressure requirements such as hopper wash down nozzles and ejector nozzle supply, if jet pumps are used. Low pressure, continuous service pumps are used to supply water seals, refractory and hopper cooling, window washing, clinker grinder and pump seals. For the latter sealing requirements automatic self-cleaning filters may be used to remove the small amount of particulate matter that may br in the stored water.

Sludge accumulations in the settling and storage tanks are returned continuously at a low rate to a dewatering bin, preferably the one into which ash is being pumped. It has been found that continuous withdrawal from these tanks during an ash handling cycle will prevent a sludge buildup in any tank bottom.

It should also be pointed out that, where vacuum for a fly ash system is hydraulically produced, water from the system storage tank can be pumped through the vacuum producer and the drainage from the vacuum producer drained back into the settling tank. Depending on the efficiency of the dust collectors ahead of the vacuum producer, some additional solids may enter the setting tank to be removed from the bottom as sludge.

The various stages of a closed-loop recirculating system are illustrated in Figures 24 to 30 inclusive. For the sake of clarity, some details have been omitted and one ash pump common to both sides of the ash hopper has been shown.

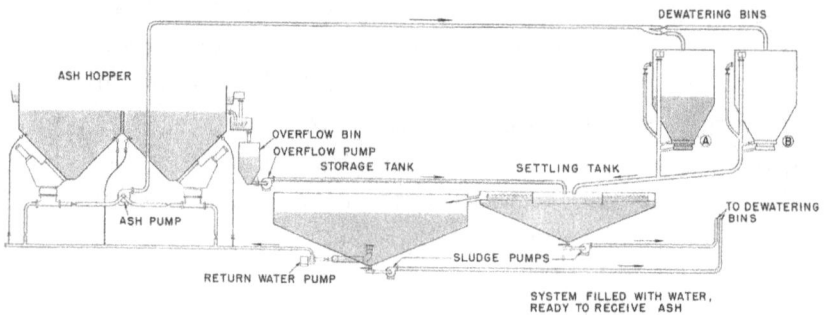

Figure 24

Initially as illustrated in Figure 24, the ash hopper is filled to its overflow line and one dewatering bin A is partially filled. Enough water remains in the storage tank to start the system in operation after the ash hopper is filled with ashes.

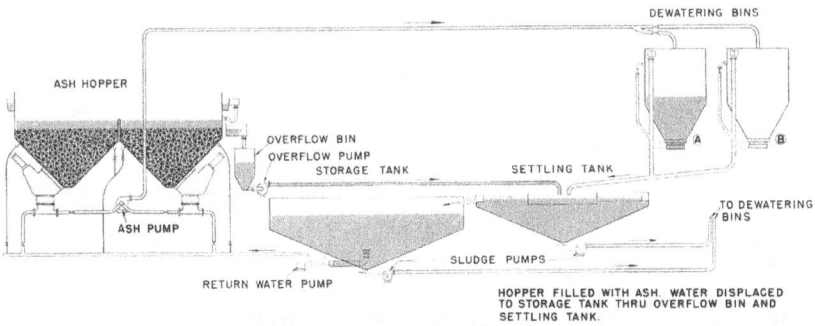

Figure 25

In the next stage, Figure 25, the ash hopper has been filled with ashes and the water displaced by them has been pumped into the setting tank and has overflowed into the storage tank. In the next step, shown in Figure 26, ash hopper cleaning is in progress with the right hand chamber partially empty. Ashes are pumped to the first dewatering bin A. As ash-water slurry enters the dewatering bin an equal amount of water overflows to the settling tank and thence to the storage tank.

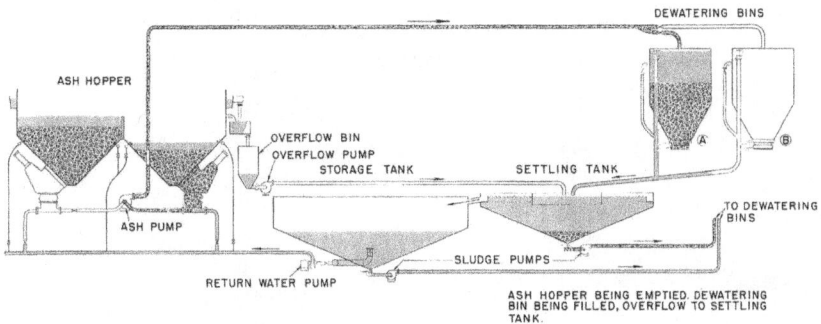

Figure 26

In Figure 27 the ash hopper has been completely emptied. All of the water volume that had been in the ash hopper is now in the storage tank ready to refill the ash hopper,

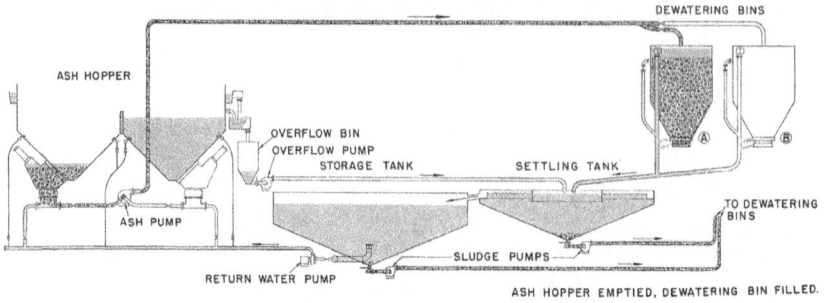

Figure 27

In Figure 28, the water trapped in the dewatering bin A is drained to the setting tank and storage tank, in Figure 29, shows where it is available for filling the second dewatering bin B as shown in Figure 30.

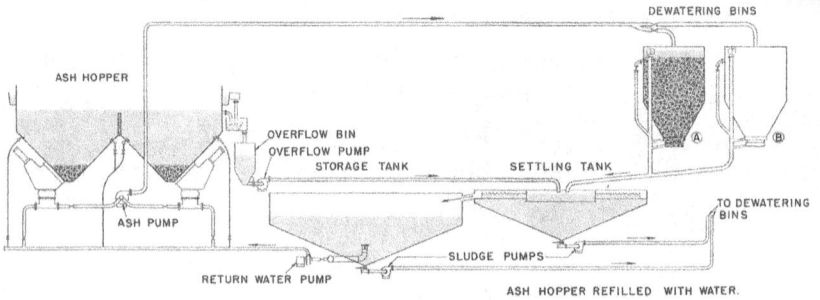

Figure 28

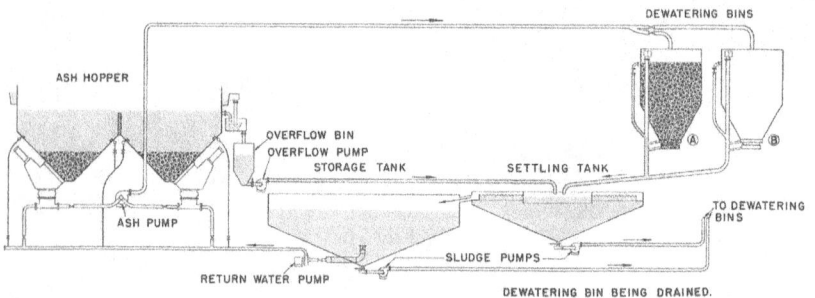

Figure 29

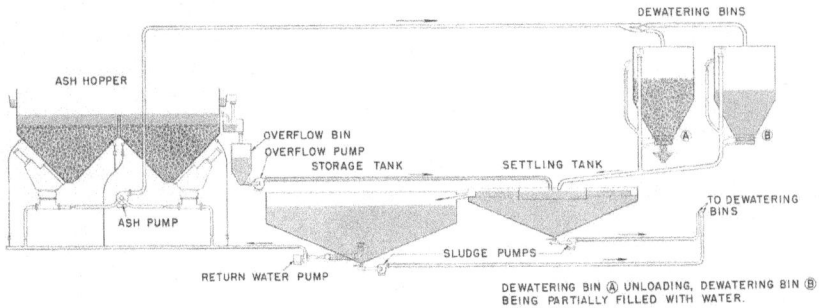

DEWATERING BINS

ASH HOPPER

OVERFLOW BIN
OVERFLOW PUMP
STORAGE TANK   SETTLING TANK

A   B

ASH PUMP

TO DEWATERING
BINS

RETURN WATER PUMP

SLUDGE PUMPS

DEWATERING BIN Ⓐ UNLOADING, DEWATERING BIN Ⓑ
BEING PARTIALLY FILLED WITH WATER.

Figure 30

It must be noted that the water volume in the setting tank remains constant while the volume in all other vessels varies during different phases of operation. An emergency by-pass can be installed between the settling tank and the storage tank to provide needed water in the event of temporary failure of outside makeup.

Makeup to the system from an outside source must be provided to restore the water lost on the bottom ash discharged from the dewatering bins. Water from this system is also used to condition fly ash before it is discharged from the silo in a separate fly ash system. This water is supplied by a pump at the storage tank having sufficient capacity and pressure to operate whatever number of dustless unloaders is required. Makeup is usually added at the storage tank.

In most cases, a closed loop recirculating system will show a marked change in the ph value of the recirculated water although this is sometimes tempered because of the amount and quality of outside makeup that must be added. Nevertheless, some ash is very high in basic compounds so that it is well to provide a monitoring system and chemical additives to maintain recirculated water as neutral as possible in order to keep pipe scaling to a minimum. Conversely, where acid conditions exist, alkali additives are required to prevent corrosion.

CHAPTER 9

Sluice System Design Calculations

In this segment we will discuss the method for calculating sluice systems for both Fly Ash and Bottom Ash. The methods of system design for water flow through a pipeline are well documented and accepted by industry. There is some doubt as to the correct factors associated with material flow in a sluice line for various materials. Sluice line velocities for various materials may vary from one manufacture to another.

For a sluice system to be calculated correctly, the designer must first recognize what type of flow will exist and the necessary velocity to maintain that flow. He should also consider the type of pipe, to be used, and what amount of wear to expect in the pipe. Fly ash will take a long time to smooth a cast iron pipe, while bottom ash may smooth cast iron pipe very fast. This being true, lower roughness factors can be used with bottom ash and cast iron pipe. For a new power plant to be initially started and come to full load operation takes a considerable time.

Elbows used in a sluice system react to flow patterns much the same as they do in pneumatic systems. But, the velocity is much lower and the segregating properties are much reduced. As a result, material does not fall out of suspension as readily and wear is not as severe. For sluice line fittings the r/d is usually 2/1 or greater. This allows for losses in terms of equivalent feet and pressure drop.

Since a slurry, of water and ash is not considered compressible, the same velocity is constant throughout a pipeline. System frictional losses for material of this type can no longer be figured as water losses times the specific gravity of the mixture. For materials of this type, the water losses must be increased by a factor of about 30% at a specific conveying rate. The designer should determine these values by a test program for each particular material until a standard for that material can be established. For most materials, operating at capacities to 100 toms/ hour, 30% is adequate. But this will not hold true for all materials. Granulated slag requires almost a 50% factor plus operating at a higher velocity.

The quantity of water to provide the correct velocity through a sluice line and the pressure to produce that flow can be provided by a head tank, a jet pump, or by a mechanical pump. If the static pressure of a head tank is adequate to provide the total required pressure, a simple makeup connection to the tank is all that is required for correct operation of the system. When a mechanical pump is used, the suction head can be added to its discharge head to provide the total head. If a jet pump is used, only a portion of the suction head is additive to the discharge head.

The data in this chapter are not intended for nor are they adequate for designing cross-country sluice systems. The designer should consult a text dealing specifically with that subject The data are sufficient for the design of most normal sluice systems found in manufacturing or electric utility plants.

Required Conveying Velocity

Fly Ash	4 to 6 ft/sec
Economizer Ash	4 to 8 ft/sec
Bottom Ash	7 to 10 ft/sec
Pyrites	6 to 12 ft/sec
Slag	6 to 12 ft/sec

Table 1

Commonly Used Values

Steel Pipe	$\epsilon = 0.00015$
Cast Iron Pipe	$\epsilon = 0.00025$
Basalt Pipe	$\epsilon = 0.0006$

Table 2

Specific Gravity

Fly Ash	2.4
Bottom Ash	1.44

Table 3

Example 1   (see Figures 31 and 32)

Fly ash is to be conveyed from a water-impounded hopper with a single outlet to a discharge pond. Power is to be produced by a Head tank, Mechanical pump, or Jet pump.

Horizontal distance = 1000 feet
Vertical rise (H_{st}) = 40 feet
Suction head (H_s) = 10 feet
Six 90° elbows (r/d = 2.0)

Tonnage rate = 50 tons/hour
Solid S.G. = 2.4
Homogeneous flow

S = specific gravity of slurry
S_l= specific gravity of liquid
S_s= specific gravity of solids
C_w= concentration of solids by weight

$$S = \frac{S_s \times S_i}{S_s + C_w (S_l - S_s)}$$

Calculations:
Required conveying velocity = 5 Feet/second (see Table 1)
Steel pipe, d = 6.07 inches or D = .506 feet
        A = 0.20 ft²
Gpm = V × A × (60) × 7.5 = 5 × 0.20 × 60 × 7.5 = 450 gpm
Concentration of solids = 35%
Specific Gravity of liquid = 1.0 $\rightarrow$

$$SG = \frac{S_s \times S_i}{S_s + C_w (S_l - S_s)}$$

$$SG = \frac{2.4 \times 1.0}{2.4 + .35 (1.0 - 2.4)} = 1.26$$

Total length = 1000 ft (horizontal) + 40 ft (vertical) = 1040 ft

(6) 90o elbows = 6 × (8) = 48 ft (see Figure 33)

Total equivalent length (T.R.L.) = 1088 ft

f = 0.0176 (see Figure 34)

$H_f = f \times (\text{T.E.L.} \times V^2 / D2g) (SG_{sl})$

Head Tank see Figure 31

$H_d = H_{st} (SG) + H_f \times SG$

$H_d = 40 (1.26) + (0.0176 \times (1088 \times (5)^2 / (0.506 \times 2 \times 32.2) \times 1.26$
$= 68.91$ ft

Mechanical Pump see Figure 31

$H_d = H_{st} (SG) + H_f - H_s$

$H_d = 40(1.26) + (0.0176 \times (1088 \times (5)^2 / (0.506 \times 2\ 32.2) - 10 = 53.13$ ft

Jet Pump see Figure 31

$H_d = H_{st} (SG) + Hf - (0.5\ H_s)$

$H_d = 40(1.26) + (0.0176 \times (1088 \times (5)^2 / (0.506 \times 2\ 32.2) - (.5 \times 10) =$
58.13

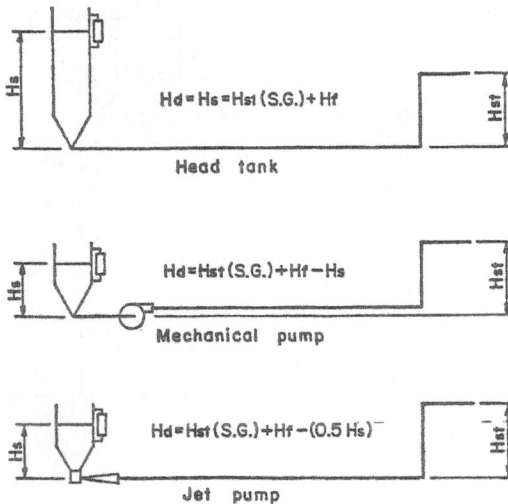

Figure 31

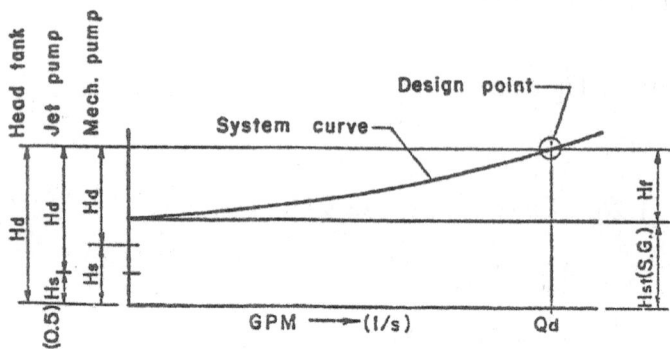

Figure 32

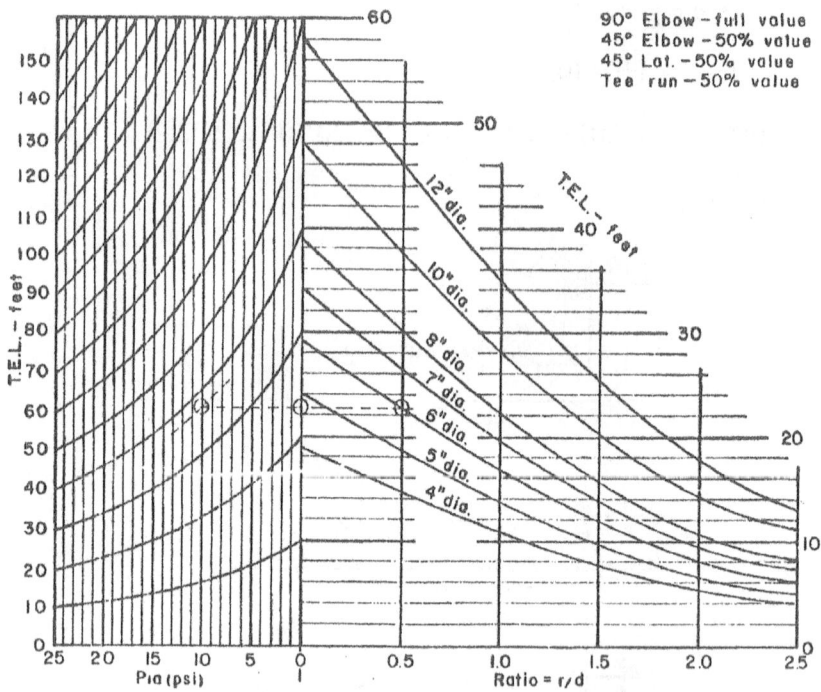

Figure 33

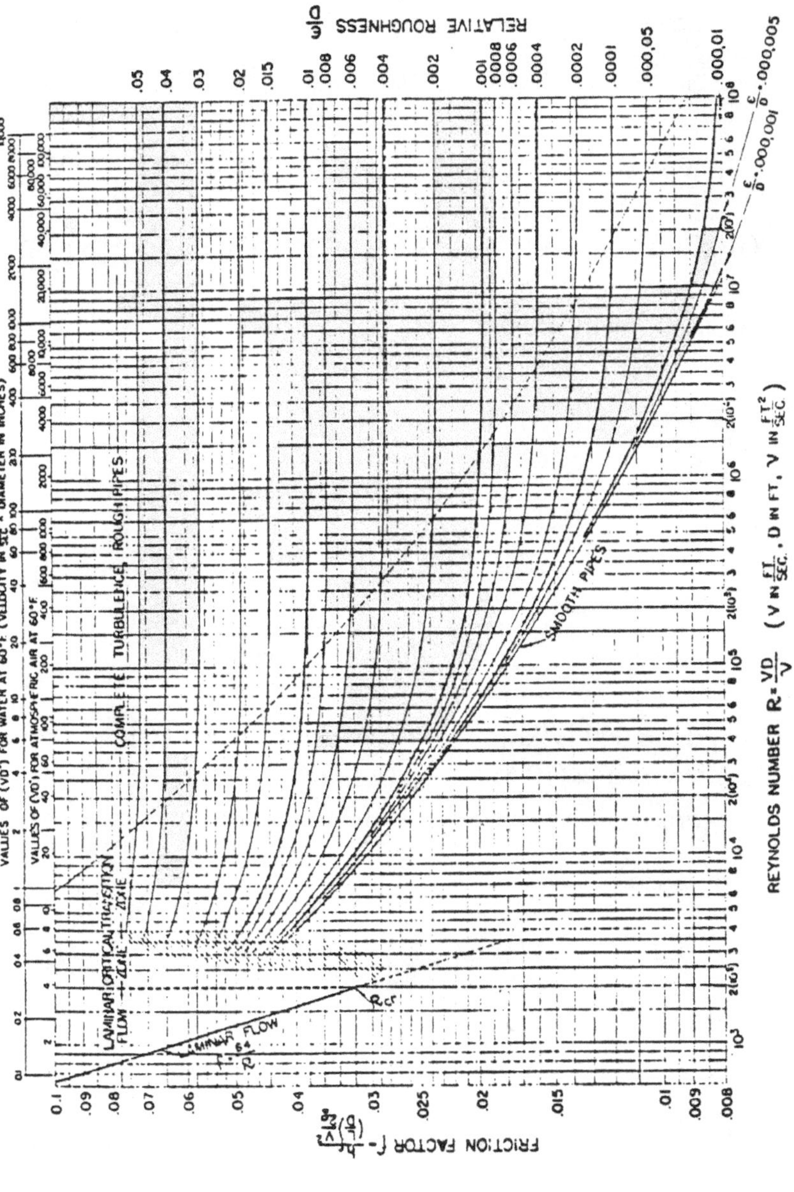
VALUES OF (VD") FOR WATER AT 60°F (VELOCITY IN FT/SEC x DIAMETER IN INCHES)
VALUES OF (VD") FOR ATMOSPHERIC AIR AT 60°F
RELATIVE ROUGHNESS ε/D
COMPLETE TURBULENCE, ROUGH PIPES
LAMINAR FLOW
SMOOTH PIPES
REYNOLDS NUMBER R = VD/ν (V IN FT/SEC., D IN FT, ν IN FT²/SEC)

Figure 34

Now if we look at the hydraulic conveying of Bottom Ash to a near by pond, as shown in Figure 35, requires additional consideration.

Required conveying velocity = 8 Feet/second (see Table 1)

Iron pipe, d = 8 inches or D = .666 feet
A = 0.348 ft^2
GPM = V × A × (60) × 7.5 = 8 × 0.348 × 60 × 7.5 = 1254
Concentration of solids = 35%
Specific Gravity of liquid = 1.0

$$SG = \frac{S_s \times S_i}{S_s + C_w (S_i - S_s)}$$

$$SG = \frac{1.44 \times 1.0}{1.44 + .35 (1.0 - 1.44)} = 1.12$$

Total length = 1000 ft (horizontal) + 40 ft (vertical) = 1040 ft
(6) 90o elbows = 6 × (10.5) = 63 ft (see Figure 33)

Total equivalent length (T.R.L.) = 1103 ft

f = 0.0202 (see Figure 35)

$H_f = f \times (T.E.L. \times V^2/ D2g) (SG_{s1})$

Head Tank see Figure 35

$H_d = H_{st} (SG) + H_f \times SG$

$H_d = 40 (1.12) + (0.0202 \times (1103 \times (8)^2/ (0.666 \times 2 \times 32.2) \times 1.12 = 74.48$ ft

Material — Bottom ash
Capacity — 100 Ton / hr
Pipe diameter — 8"

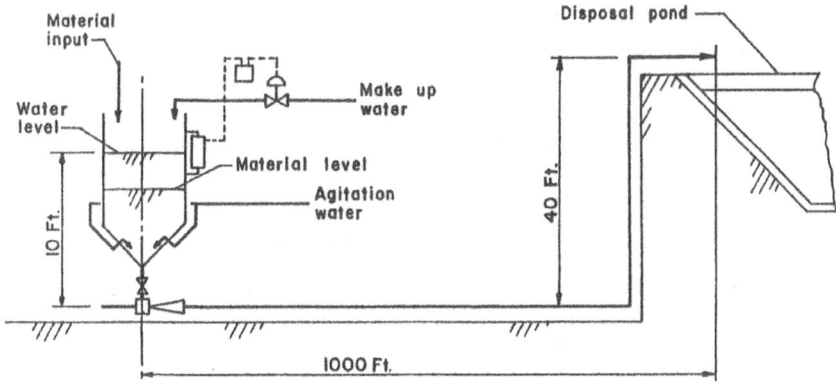

Figure 35

If the bottom ash is to be conveyed to a dewatering bin, the total discharge head required is greater than a jet pump can deliver without excessive nozzle pressure. It still may be desirable to use a jet pump at the bottom ash hopper to minimize controls and allow more operator flexibility. A transfer tank could be used, with a material handling pump to convey from there to the dewatering bin. See Figure 36

1. Bottom ash hopper water level will be lowered each operation.

2. Material will be pulverized fuel ash (see table 1 for velocity requirements)

3. Conveying rate is 100 tons per hour.

4. Material density is 45 lbs/ft³.

5. Solid specific gravity is 1.44

6. See Figure 36 for system layout

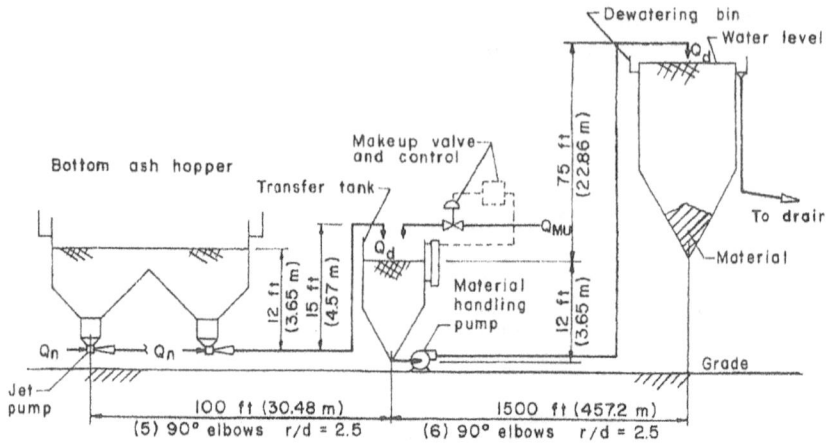

Figure 36

One of the main criteria in selecting the diameter of the sluice line should be the total time necessary to lower the water level in the bottom ash hopper. That is, the suction gpm, $Q_s = Q_d - (Q_n + Q_{re} + Q_{cr} + Q_w)$, should be great enough to minimize the time required to lower the water level. To accomplish this, it is generally agreed that $Q_n = 0.65$ to $0.85\ Q_d$. For this arrangement, the length of line from the bottom ash hopper to the transfer tank is relatively short and the chances of the line plugging are greatly reduced. Thus, relaxing of the minimum nozzle gpm requirement. For a moderate-sized boiler, the following can be assumed for initial design.

$Q_{re} = 350$ gpm (refractory cooling @ 2 gpm / foot of weir)
$Q_{cr} = 20$ gpm (clinker grinder seal water)
$Q_w = 30$ gpm (window cooling water)
$Q_n = 800$ to 1200 gpm (jet pump nozzle water)
$Q_d = 2000$ to 2400 gpm (design flow)
Q_d for 8 inch diameter pipe = 1250 gpm at 8 feet / second
Q_d for 10 inch diameter pipe = 1960 gpm at 8 feet / second
Q_d for 12 inch diameter pipe = 2825 gpm at 8 feet / second

For a 8" diameter pipe, Q_s = 1250 – (800 + 350 + 20 + 30) = 1250 - (1200) = 50 gpn

Therefore an 8 inch diameter pipe line does not meet the requirements.

For a 10" diameter pipe, Q_s = 1960 – (800 + 350 + 20 + 30) = 1960 – (1200) = 760 gpm

For a 12" diameter pipe Q_s = 2825 – (800 + 350 + 30) = 2825 – (1200) = 1625 gpm

As can be seen, the use of the 10" pipe would be acceptable. The listed gpm, Q_d, are for design with material. Greater quantities will be handled by the jet pump for water flow only. See Figure 37.

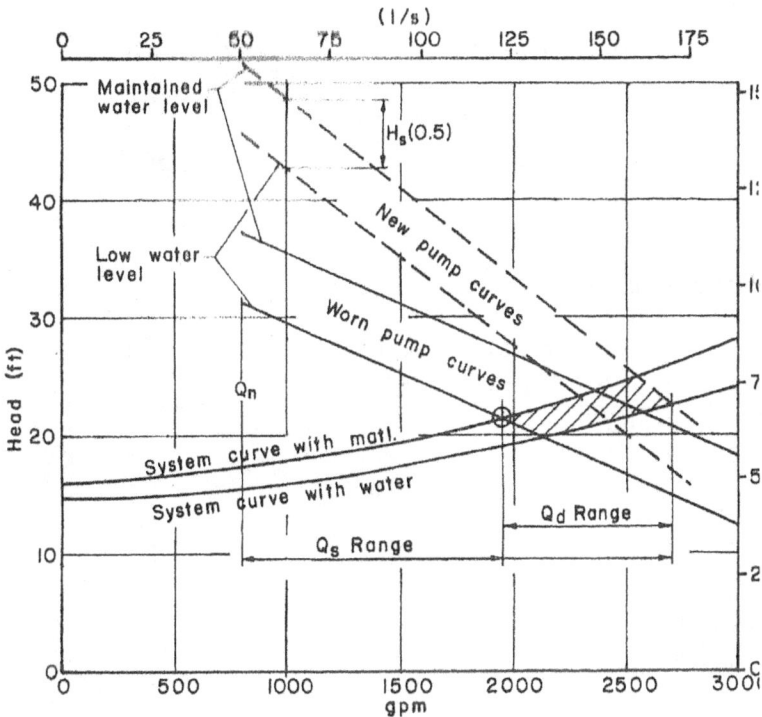

Figure 37

By using a 10 inch line from the bottom ash hopper to the transfer tank allows a 12 inch line to be used from the transfer tank to the dewatering bin.

The total amount of slurry handled by the jet pump will vary from a minimum at maximum loading to a maximum with clear water. This variation will change as the combining tube, venture throat, wears from new to worn. Thus the quantity of water or slurry discharged to the transfer rank will vary greatly. The material handling pump must be capable of handling the maximum gpm discharged from the jet pump. Since this is a variable, a makeup supply controlled at a set level must be furnished to the transfer tank. Wear of the material handling pump should be accounted for, but this will normally be much slower than the wear experienced at the jet pump.

Calculations
Required conveying velocity = 8 feet / second (see table 1)
10 inch cast iron pipe, d = 10 inch
$$A = 0.545 \text{ feet}^2$$
Required gpm = Q_d = A × V = (0.545) (60 sec./min) (7.5 gal./ft³) (8 ft/sec) = 1950

Bottom ash hopper to transfer tank:

Total length = 100 ft (horizontal) + 15 ft (vertical) = 115 ft
(5) 900 elbows = 5 × (10.5) = 53 ft (see Figure 33)
Total equivalent length (T.R.L.) = 168 ft

C = 0.00085 (see Figure 38)
C/D = 0.00085 (12/10) = 0.00102
V_d = 8(10) = 80
f = 0.0202 (see Figure 34)
F = 1.3 (material loss factor)

H_d (material) = $H_{st}(SG) + H_f(F) + 0.5H_s$
= $H_{st}(SG) + (f(TEL)V^2/2g\ D)F + 0.5H_s$

$= 15(1.12) + (0.0202\ (168)(8)^2\ /\ (2 \times 32.2\ x(10/12)\ )) \times (1.3) + 0.5(0) = 21.10$ ft

Only the calculations for the design point are shown. The designer may wish to calculate and check the other curves.

Transfer tank to dewatering bin:

Total length = 1500 ft (horizontal) + 87 ft (vertical) = 1587 ft
(6) 900 elbows = 6 × (14) = 84 ft

Total equivalent length (T.E.L.) = 1671 ft

Sluice System Calculations
    Gpm = 2825
    ε = 0.00085
    ε/D = 0.00085 (12/12) = 0.00085
    V_s = 8 (12) = 96
    f = 0.02
    F = 1.3 (material loss factor)

$$SG = \frac{S_s \times S_i}{S_s + C_w\ (S_i - S_s)}$$

$$SG = \frac{1.44 \times 1.0}{1.44 + .35\ (1.0 - 1.44)} = 1.12$$

H_d (material) $= H_{st}\ (SG) + (f(TEL)\ V^2 /\ (2gD)\ F + H_s$

$= 87(1.12) + (0.02(1671)(8)^2\ /\ (2 \times 32.2(12/12)) \times 1.3 + (-12)\ 1 = 123.27$ ft
H_d (water) $= 87(1.0) + (0.02\ (1671)(8)^2\ /\ (2 \times 32.2)(12/12)) \times 1.0 = 108.21$ ft

Calculations for the design gpm for water and for 100 tons/hour are shown. The designer may wish to calculate and check the other curves. The amount of wear allowance varies widely among designers but can be assumed as about 5 to 10% of the total discharge head

$H_{d \text{ (design point)}}$ (1.05 to 1.1).

With reference to Figures 37 and 39, it can be seen that the material handling pump has the capability to handle more water and/or slurry than the jet pump can provide. It is then necessary to add makeup to the transfer tank to satisfy the material handling pump.

Both systems could have been designed on any gpm over the minimum value of 8 ft/sec. to achieve any desired balance. It is desirable to lower the water level in the bottom ash hopper as rapidly as possible. Thus the suction gpm to the jet pump could be increased by raising the nozzle pressure and/or gpm. This, in effect, would decrease the total system operating time. The designer should weight the various alternatives and decide which condition best suits the particular system.

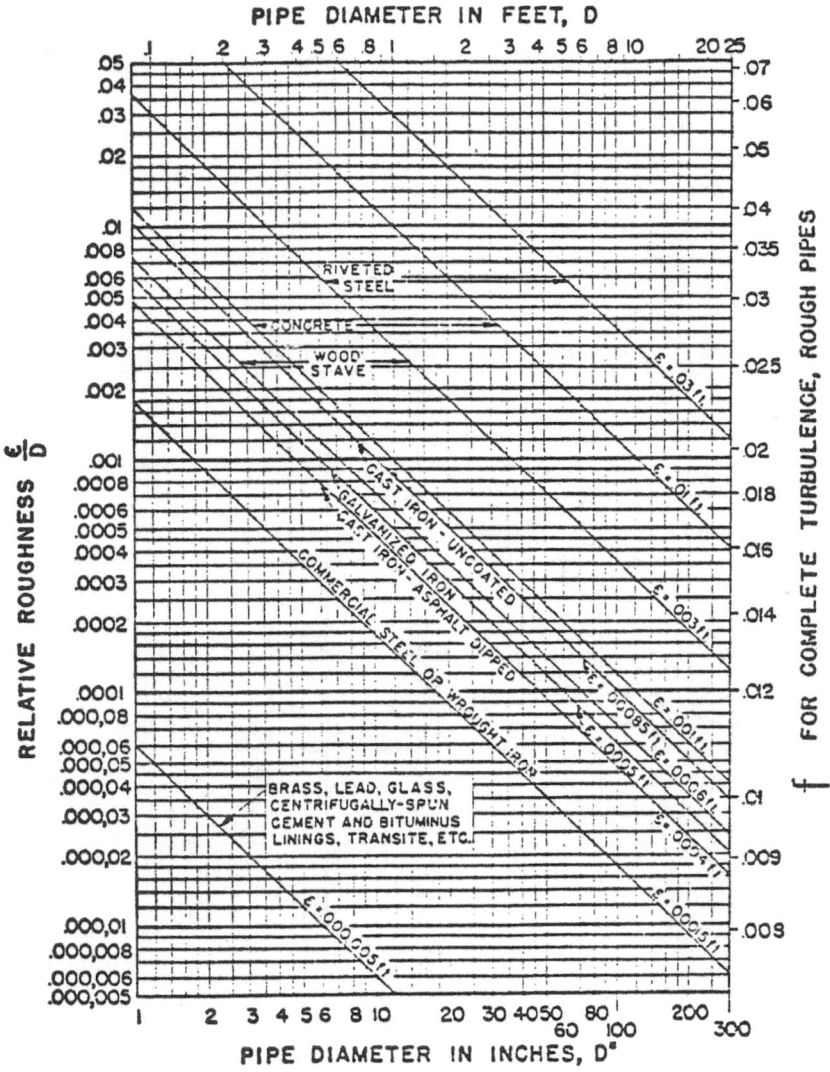

Figure 38

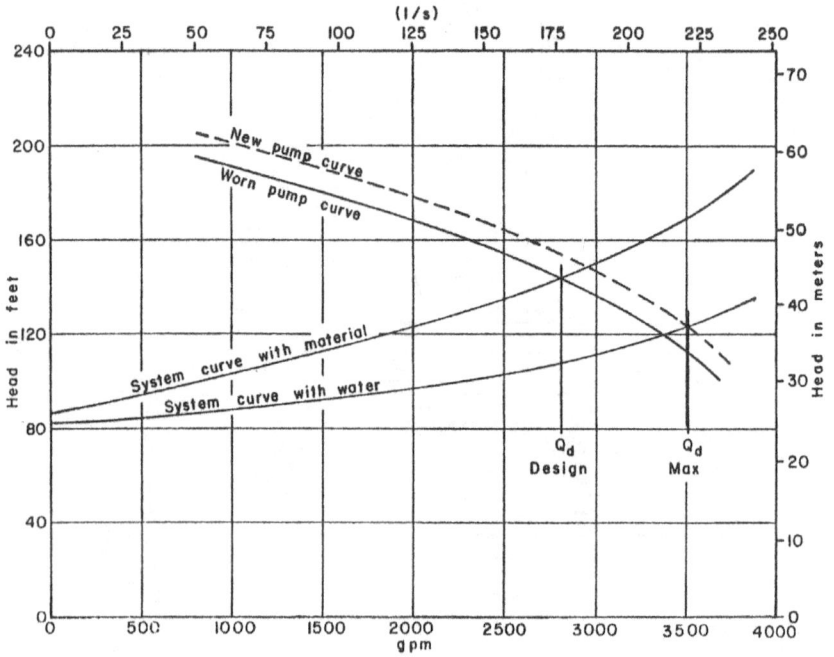

Figure 39

CHAPTER 10

Introduction to Pneumatic Conveying

There are three types of pneumatic conveying of solids. See Figure 40. Negative Pressure (Vacuum Systems), Positive Pressure, and Air Assist Gravity Flow Systems, and a definition of these systems in Figure 41.

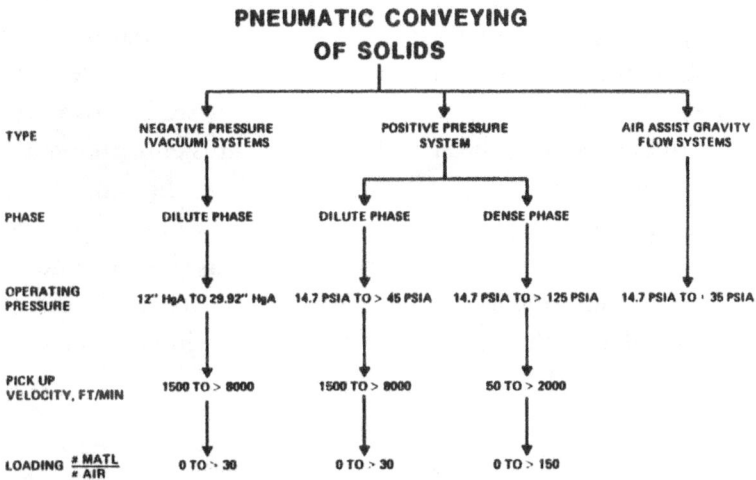

PNEUMATIC CONVEYING OF SOLIDS

TYPE	NEGATIVE PRESSURE (VACUUM) SYSTEMS	POSITIVE PRESSURE SYSTEM		AIR ASSIST GRAVITY FLOW SYSTEMS
PHASE	DILUTE PHASE	DILUTE PHASE	DENSE PHASE	
OPERATING PRESSURE	12" HgA TO 29.92" HgA	14.7 PSIA TO > 45 PSIA	14.7 PSIA TO > 125 PSIA	14.7 PSIA TO · 35 PSIA
PICK UP VELOCITY, FT/MIN	1500 TO > 8000	1500 TO > 8000	50 TO > 2000	
LOADING $\frac{\text{\# MATL}}{\text{\# AIR}}$	0 TO · 30	0 TO · 30	0 TO > 150	

Figure 40

SYSTEM DEFINITIONS

DILUTE PHASE

A SYSTEM WHERE:
THE MATERIAL IS CARRIED BY AN AIR STEAM OF SUFFICIENT VELOCITY TO ENTRAIN AND RE-ENTRAIN IT FOR A DISTANCE DEPENDENT ON THE AVAILABLE PRESSURE.

DENSE PHASE

A SYSTEM WHERE:
THE MATERIAL IS PUSHED THROUGH A PIPELINE AS A SLUG OR AS A FLUIDIZED SLUG FOR A DISTANCE DEPENDENT ON THE AVAILABLE PRESSURE. ADDITIONAL AIR MAY BE ADDED ALONG THE LENGTH OF THE PIPELINE TO RE-FLUIDIZE THE MATERIAL.

AIR ASSIST GRAVITY FLOW

A SYSTEM WHERE:
AIR IS PASSED THROUGH A POROUS MEMBRANE OF LOW PERMEABILITY, FILLING THE VOIDS OF THE MATERIAL DIRECTLY ABOVE IT AT A LOW PRESSURE, CHANGING THE ANGLE OF REPOSE OF MATERIAL ON MATERIAL AND CAUSING IT TO FLOW BY GRAVITY.

Figure 41

In dilute phase systems, there are three types of material flow that are considered. Horizontal, vertical and incline flow. Figure 42 defines these conditions. Material can be conveyed vertically upward at a lower velocity than horizontally because the velocity component is only opposed by the gravity and drag components. In horizontal flow, the same forces are present, but gravity is at ninety degrees to the other forces. This requires higher velocity for re-entrainment. Inclined flow is similar to horizontal flow but has an additional slip component to overcome. We discourage inclined lines due to the problems encountered trying to overcome this slip component.

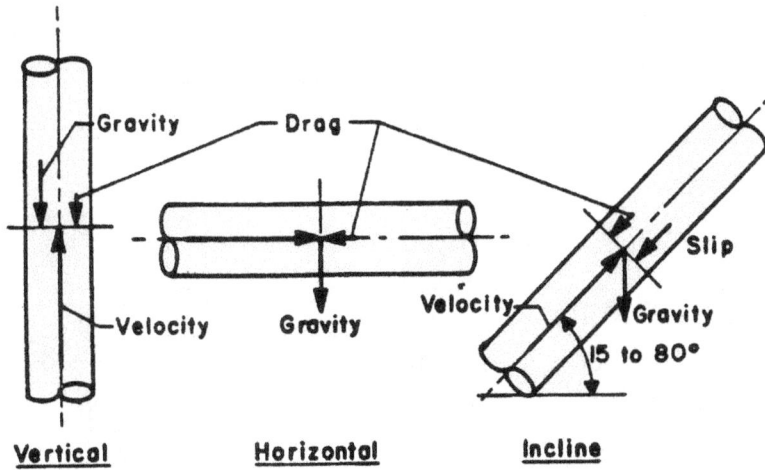

Figure 42

Each material conveyed has a minimum pick-up velocity that must be maintained or exceeded. In dilute phase work, the only velocity referred to is the air velocity, and it is as though the material occupied no space. There also is no one velocity value for any one material. It is dependent on material shape, size, density, moisture content and chemical analysis. Published values vary mainly because they were set by individual observers, each with a different idea of what was required. Values shown in Table 4 are realistic; however, they cover the full range of values used by various manufacturers.

Suggested pick up velocity

Type of Material	Feet / minute
Bottom Ash	
P.C. Boiler	3800 to 5200
Chain Grate Stoker	4200 to 5600
Underfeed Stoker	4700 to 6200
Fly Ash	1800 to 4600
Pyrites	5300 to 7000

Cement Dust	3500 to 4600
Hydrated Lime	3800 to 5000
Pebble Lime	3800 to 5000

Table 4

Commonly used values for Fly Ash (feet/minute)

U.C. (United Conveyor)	3800
A.S.H. (Allen Sherman Hoff)	1800 to 3800
C.E. (Combustion Engineering)	3800
D.S. (Detroit Stoker)	3000
National (National Conveyor)	4600

Table 5

The A-S-H value of 1800 ft/min. was the minimum used for negative pressure systems burning Eastern coal while the 2300 ft/min. value was used for Western coals. The present A-S-H Standards call for 2000 – 2200 ft/min. for Eastern coal, 2500 – 3200 ft/min. for Western coal, and 2700 ft/min. for lignite. These are minimum velocities with A-S-H using higher velocities as required for a particular coal seam.

Each company figures fitting losses differently and each of them look at pipe diameter and combinations of pipe diameters differently. Each company has a method of calculating or an equation that is relentlessly guarded.

Starting velocities as listed in Table 5 are good for all negative pressure systems. They can be higher for use with positive pressure systems due to the greater air density at operating pressure. However, A-S-H has historically used higher velocities for pressure systems. If the resulting velocities are too low, slug flow or a plugged system will result. Either of these conditions should be avoided at all cost.

Slug flow is not limited to low velocities but can be induced by the use of some elbow configurations, see Figure 43.

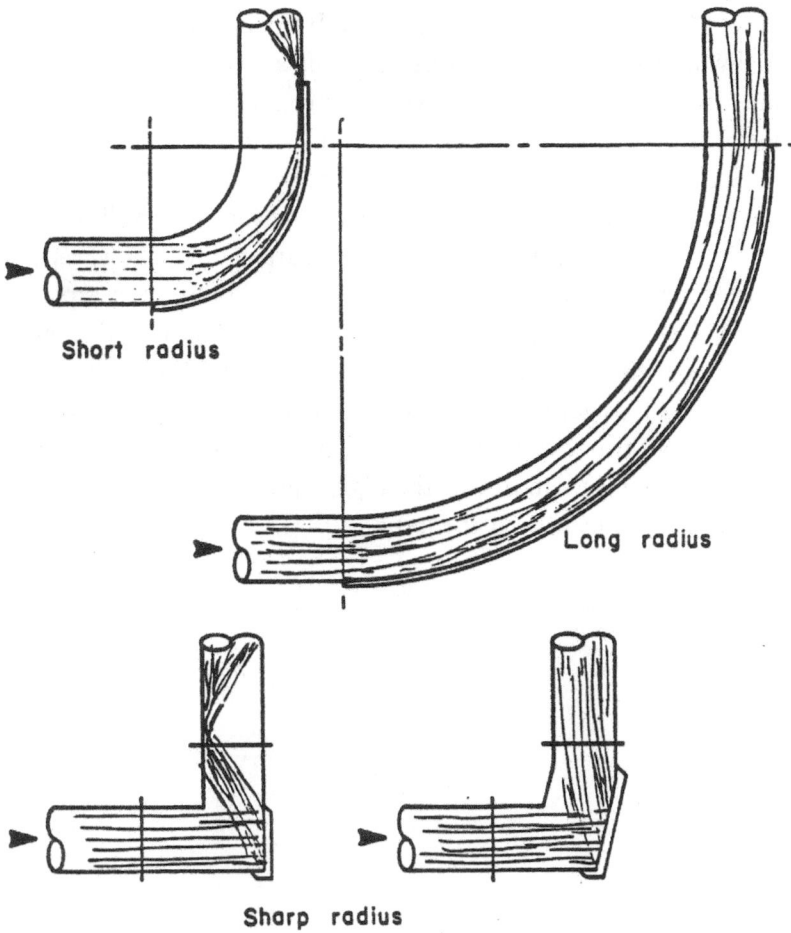

Short radius

Long radius

Sharp radius

Figure 43

An elbow with a radius acts like a centrifugal separator and tends to force all material out of suspension. Depending on velocity, radius, concentration of pounds of material per pound of air, elbow orientation, and other factors, the material may drop out of suspension at the discharge of the elbow. If the elbow orientation is horizontal, the drop out effect may be minimal or may cause considerable drifting beyond the elbow. This drifting can eventually cause slug flow to occur. T he worst place for this type of elbow would be located is at the base of a riser.

Here, as the material is separated out of the air stream, it moves upward very slowly, hugging the face of the elbow. If the conditions are right, the material will drop into the air stream, causing a slug to move through the pipeline. There are two methods commonly used to guard against this phenomenon. First, use elbows of extra long radius, with a ratio of L/D of 4/1. Secondly, sharp radius elbows can be used. This in effect absorbs the energy as impact forces with minimal abrasion. It is well to include a short wear section at the outlet of each sharp radius elbow, especially when steel pipe is used for the conveying line. Elbows with replaceable wear backs have been used for years and are still used for negative pressure systems, but their economical advantage no longer exists. They should never be used for negative pressure system handling hot materials, due to softening the wear backs upon impact.. They should also not be used in a positive pressure system due to the higher pressure, and the possibility of air and material leakage at the joint.

CHAPTER 11

Negative Pressure Pneumatic Conveying Systems

Negative pressure systems, for the most part, can be considered as dilute phase. By definition, in a dilute phase system, the material is carried by an airstream of sufficient velocity over a distance dependent on the pressure available. If the material is fed into the pipeline too fast, the pressure will rise and the velocity will drop with a resultant plugged line. Thus, fly ash intakes are furnished to regulate the amount of material into the pipeline.

There are two commonly used types of intakes on the market:

The swing gate types that several manufacturer use, which uses a disc type gate that swings open and is between the intake and the conveying line. See Figure 44. Full load control is accomplished by opening and closing the gate with full conveying air supplied through the air intake. This system depends on the continuing operation of the materials handing valve opening and closing as to not overload the conveying line.

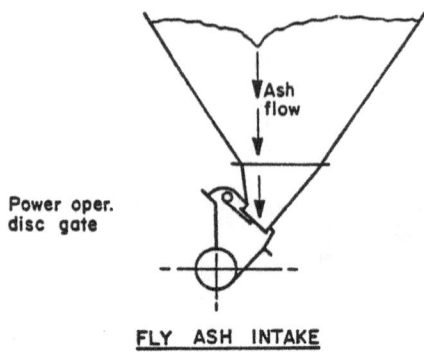

FLY ASH INTAKE

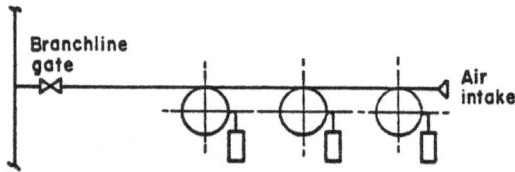

TYPICAL BRANCHLINE

Figure 44

The other version which introduces most of the conveying air through the intake itself. See Figure 45. The intake, which in effect is a 90 degree elbow, isolates the hopper from the conveying line by virtue of its shut off gate being in the horizontal line. The air intake at the end of the conveying line is normally restricted and requires a fairly large pressure drop to admit conveying air. Thus, a scavenger valve located downstream of the ash intakes and before any riser had been used in some cases for full load regulation, in resent years, this valve has been eliminated. It can be shown and proven that after the initial discharge of material from a hopper that almost all conveying air, or gas comes from the hopper and not an outside source regardless of whose intake is used.

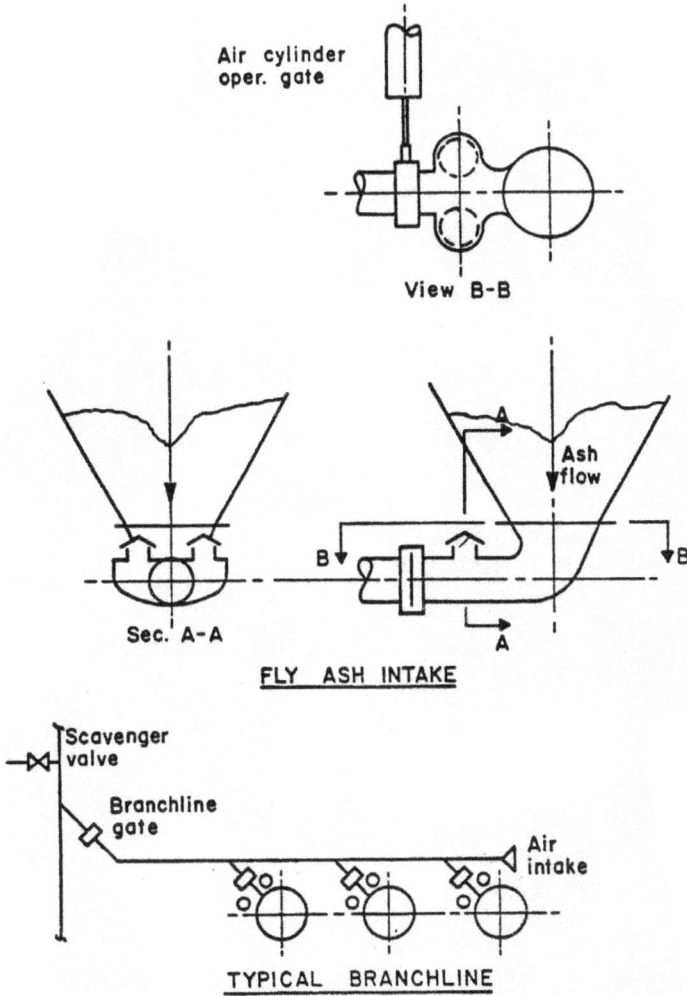

Air cylinder
oper. gate

View B-B

Ash
flow

Sec. A-A

B

B

A

A

FLY ASH INTAKE

Scavenger
valve

Branchline
gate

Air
intake

TYPICAL BRANCHLINE

Figure 45

Air intakes, for bottom ash systems used in negative pressure systems, are virtually the same for all manufacturers. They essentially are a tee connection which allows gravity feed. Most bottom ash systems are operated manually. There are some that are operated automatically, and they follow along the same lines as a fly ash system. Figure 46 depicts the general arrangement.

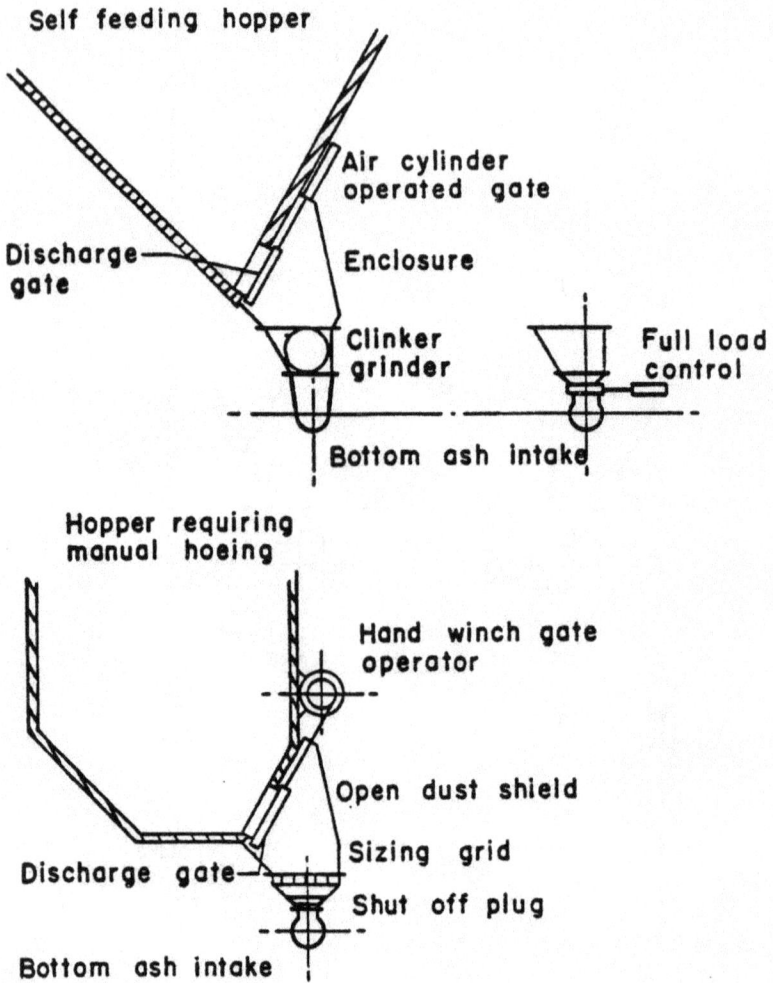

Self feeding hopper

Air cylinder
operated gate

Discharge gate

Enclosure

Clinker grinder

Full load control

Bottom ash intake

Hopper requiring manual hoeing

Hand winch gate operator

Open dust shield

Sizing grid

Discharge gate

Shut off plug

Bottom ash intake

Figure 46

For an automatic system, the ash hoper has to be shaped so it can empty completely by gravity flow. A clinker grinder is some times incorporated, to size the clinkers to a point that won't plug the system down stream.

The intake of a manually operated system can contain a sizing grid or a clinker grinder about it as well. When the grid is used, an operator must physically break up the clinkers to pass through the grid.

Material Collecting and Separating Equipment

To separate the dry material from the gas/airstream, and deposit it into a silo or transfer vessel, requires the use of separating equipment. Usually this type of equipment is centrifugal separators. The size and type of separators required for any particular installation depends on the type of material being conveyed, the percent fines, and the tonnage. Each manufacturer supplies the type equipment that suites the needs from the various separators he stocks. Most systems will use a primary separator of one and one half to six feet in diameter, with various diameter inlets and outlets. Primary collectors are generally about 85% efficient due to the heavy grain loading. Secondary separators will range from ten inches to four feet in diameter with an overall efficiency of about 85%. If greater efficiencies are required, a bag filter or some type of water scrubber is used. Bag filters generally range in efficiencies of about ninety-nine point nine percent (99.9%).

When hot material is conveyed, the gate sealing must be capable of withstanding the temperature and extreme care must be used if bag filters are still to be used. Generally, it might be better to use a water operated scrubber in place of the bag filter. Should one of the hot particles being conveyed reach a bag in the filter, a small hole will be burnt in the bag and its efficiency will decrease rapidly.

Special care should be taken when using a centrifugal separator with a vacuum system, whether it is a continuous or intermittent system. On some system suppliers, the separator generally contains a cone or deflector plate located centrally and well below the inlet connection. Full load material level can normally be considered as six to nine inches below the cone. Other type separators do not have cones or deflector plates.

The full load level is considered about eighteen to twenty four inches below the baffle plate which extends down from the top plate.

For all practical purposes, the separator operates at system design vacuum. Thus, the material is collected in a rarified atmosphere. In this state, fine or powered material will not flow out of the separator regardless of the discharge gate diameter. It is therefore necessary to have aeration holes in the discharge gate to allow air to fluidize the material in some manner. It is better to use many small holes than it is to use a few large holes, say one sixteenth to one eighth inch diameter. The discharge gate should fit tight without leaking. A leaking gate will not give the same results as the use of many small holes. See Figure 47 and 48.

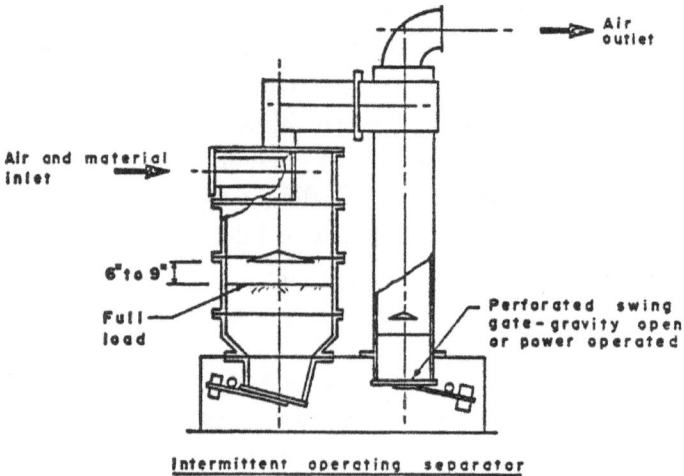

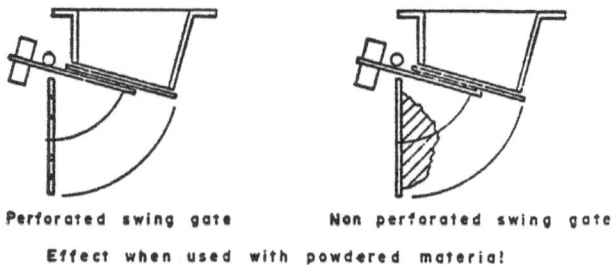

Figure 47

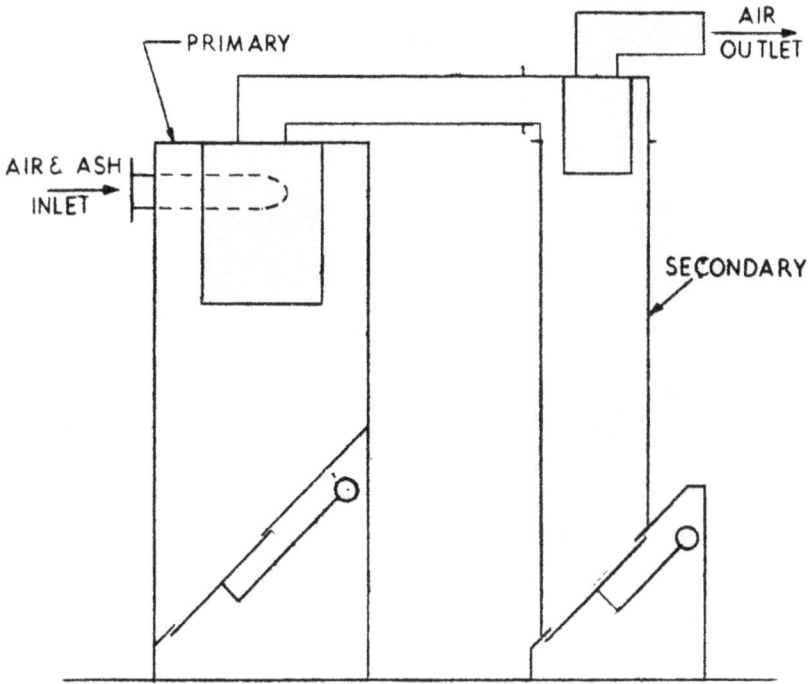

AIR
OUTLET

PRIMARY

AIR & ASH
INLET

SECONDARY

A-S-H INTERMITTENT OPERATING SEPARATORS

INTERMITTMENT OPERATING SEPARATORS
Figure 48

When a bag filter is used, it generally is in series with a primary and sometimes a secondary collector. A filter used in this service must be designed to operate from atmospheric pressure to full maximum vacuum. The Filter is always of the reverse pulse jet cleaning type. See Figure 49 for arrangement of a combination primary and secondary collector for continuous service. Figure 50 shows a typical jet pulse bag filter.

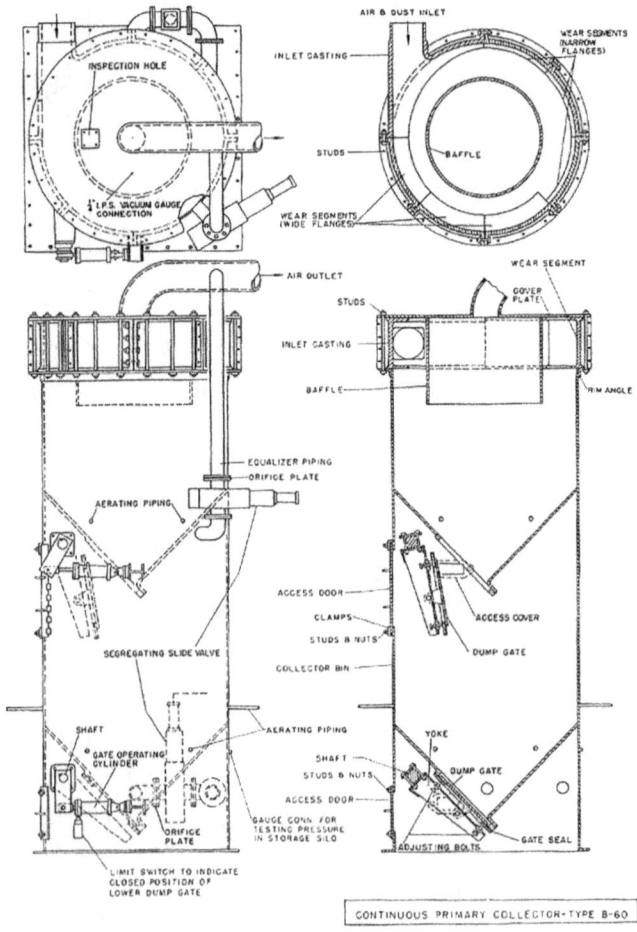

Figure 49

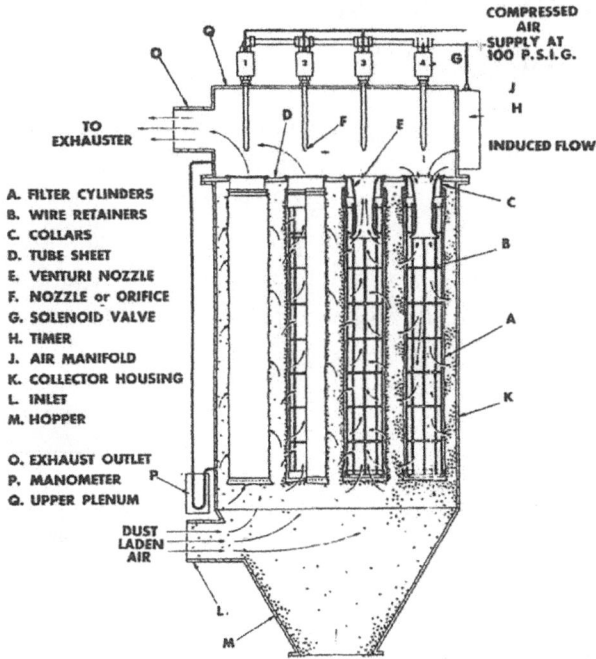

A. FILTER CYLINDERS
B. WIRE RETAINERS
C. COLLARS
D. TUBE SHEET
E. VENTURI NOZZLE
F. NOZZLE or ORIFICE
G. SOLENOID VALVE
H. TIMER
J. AIR MANIFOLD
K. COLLECTOR HOUSING
L. INLET
M. HOPPER

O. EXHAUST OUTLET
P. MANOMETER
Q. UPPER PLENUM

Figure 50

Control System

Negative pressure pneumatic conveying systems are furnished with several different types of material intakes, various combinations of separating equipment, and three different types of vacuum producers. As a result, there is no one control system that will satisfy all the needs. A good understanding of how a system works will help in understanding the various control circuits. There are three basic vacuum systems, the continuous operating system, no separating equipment, intermittent operating systems, and continuous system with separating equipment. Each of the three will be covered separately.

CONTINUOUS OPERATING SYSTEM – NO SEPARATING EQUIPMENT

Figure 51 shows this form of system, which uses no separating equipment, and merely conveys the material directly through a water vacuum producer. The dry material and air, mix with the water and flow by gravity to an area capable of holding both the material and water. If the sluice line from the vacuum producer is long, an air separator is required to separate the air from the slurry to prevent possible line pluggage due to air pockets.

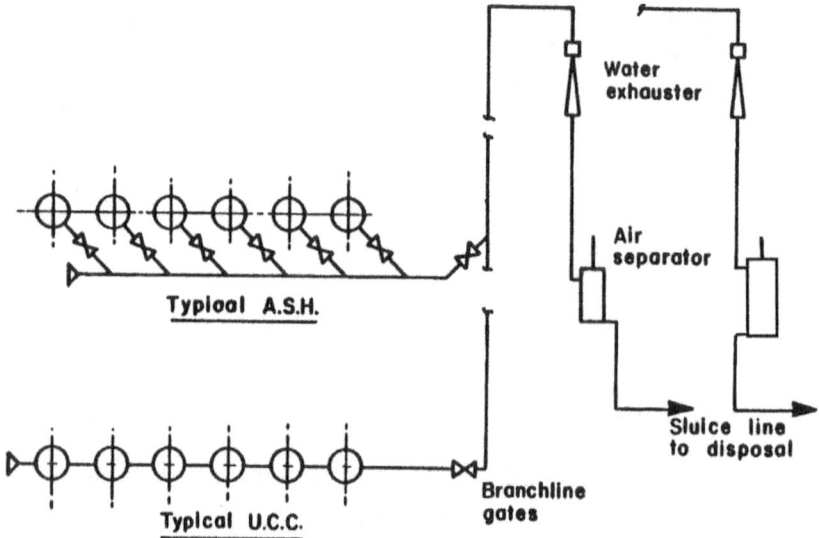

Figure 51

This type of system is one of the oldest and most reliable, but unfortunately its use is now limited by ground water pollution laws. Not only does the disposal area have to prevent leaching of elements into the ground, the run-off must be of a quality that can be returned to a stream or river.

Full load control whether by the A-S-H method which uses a check vale at the end of the conveying line to control the amount of air entering the conveying line, or the U.C.C method which has the materials handling valve gate opening and closing, gives the same results. Full load is considered to be the system design vacuum as measured at the inlet to the vacuum producer. See Figure 52.

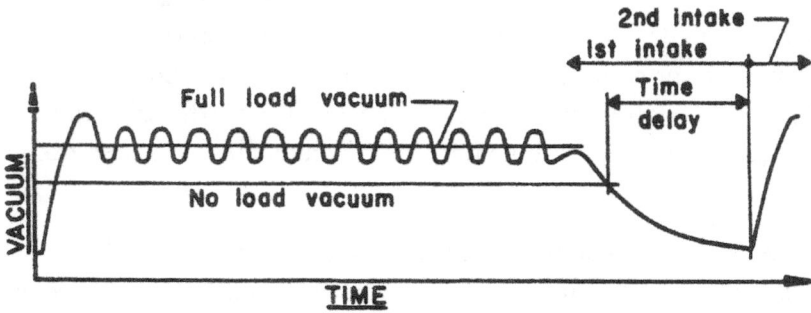

Figure 52

The A-S-H design philosophy is to use higher vacuum level and lower velocities versus the U.C.C. method of lower vacuum level and higher velocity.

The sequence of operation is to pull the ash from the farthest branch line first, and the farthest hopper in that branch line first. With this sequence of operation, the fly ash in each hopper will aid in preventing air flow from that hopper to enter the conveying line, until the valve is opened. If the sequence were reversed, air could be pulled from an empty hopper, since the gate and seat are not bubble tight, and it would be in front of the ash that was to be conveyed. The conveying media (air) has to be behind the material in order to pick it up and convey it to the disposal point. Also, while air was being pulled through this valve, any small amount of ash that may still be present will also be pulled through the small gap and cause wear due to the wire drawing effect.

Transfer from one hopper to the next is treated the same by most manufacturers. When a hopper is empty of material, or due to a rat hole through the material, the vacuum will drop and approaches that of air flow alone. A vacuum switch set for no load (or vacuum low) will signal the transfer mechanism. In some installations, there are hopper vibrators. When these are used, first they would be mounted on the outside of the hopper. Internal vibrator systems tend to be very ineffective and detrimental, since they could interfere with material flowing out of the hopper. The control scheme for the vibrator control is at a vacuum level close to the vacuum low level. As the hopper empties and the vacuum level starts to drop toward the vacuum low point another vacuum switch signals vibrator control. If there is still material in the hopper when the vibrator is energized, the flowing material will cause the vacuum level to rise again. This will continue until all the material is evacuated from the hopper.

The vacuum low switch is normally in series with a time delay set for a few seconds. If the vacuum remains at this level for this time, the sequence switch is energized which will transfer operation from one hopper to the next or to shut down. The sequence switch also controls the branch line segregating valves and the water valve.

A vacuum breaker valve is located as close to the hydraulic vacuum producer as possible. This valve is a normally open valve. When the system is energized, this valve closes to develop the vacuum required to convey ash. At shutdown, the vacuum breaker opens immediately to prevent the vacuum in the system from pulling water back into the conveying line.

INTERMITTENT OPERATING SYSTEMS

The basic operation of the intermittent operating system is the same as the continuous operating system as far as the conveying line is concerned. They differ in that the intermittent operating system utilizes material collection and discharge equipment along with a storage silo. The term intermittent refers to system operation where material is conveyed and collected in a separator for a period of time. At the completion of the conveying portion of the cycle, all material flow is stopped by closing the material handling valve and the conveying line is purged. The collected material in the collector is discharged to the silo, after which conveying and collecting is resumed. The total time cycle, usually is from thirty seconds to two minutes, is repeatable as long as there is material to convey and the system remains in operation. See Figure 53.

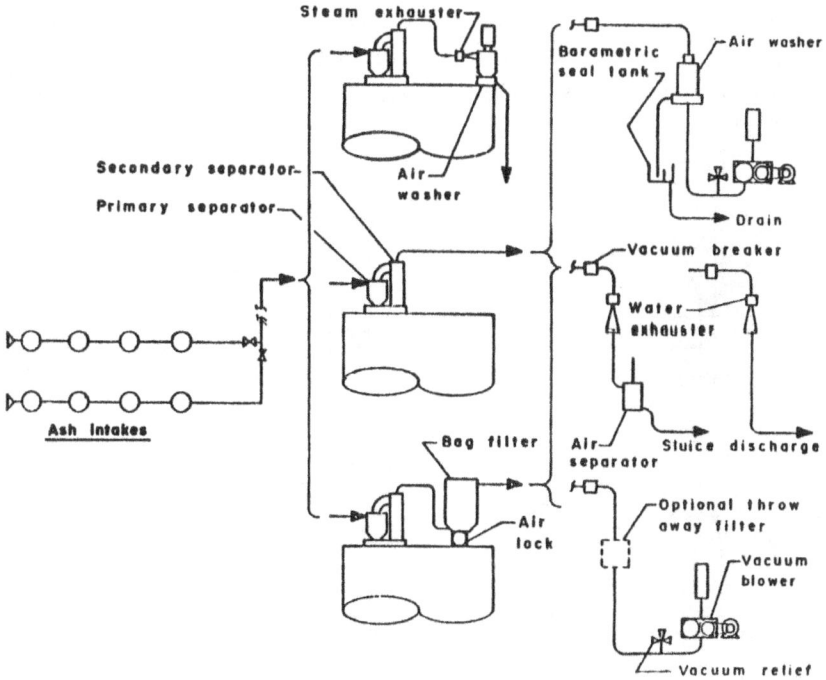

Figure 53

Since the material is alternately collected in a separator and then discharged from it, the gross conveying rate is defined as the maximum instantaneous conveying rate and differs from the average conveying rate. The average conveying rate is defined as the overall conveying rate or guaranteed capacity.

Gross capacity = net capacity/cycle efficiency

Net capacity = guaranteed capacity (no load factor)

The no load is usually equal to ten percent (1.1). It is an allowance for the lost time due to "No Load" transfer between hoppers and to shut down.

See Figure 54. The (x) and (y) figures represent time to re-establish air flow and time to purge the line of material after a materials' handling valve has closed. The value of (x) is usually two seconds while the value of (y) can vary from three to seven seconds.

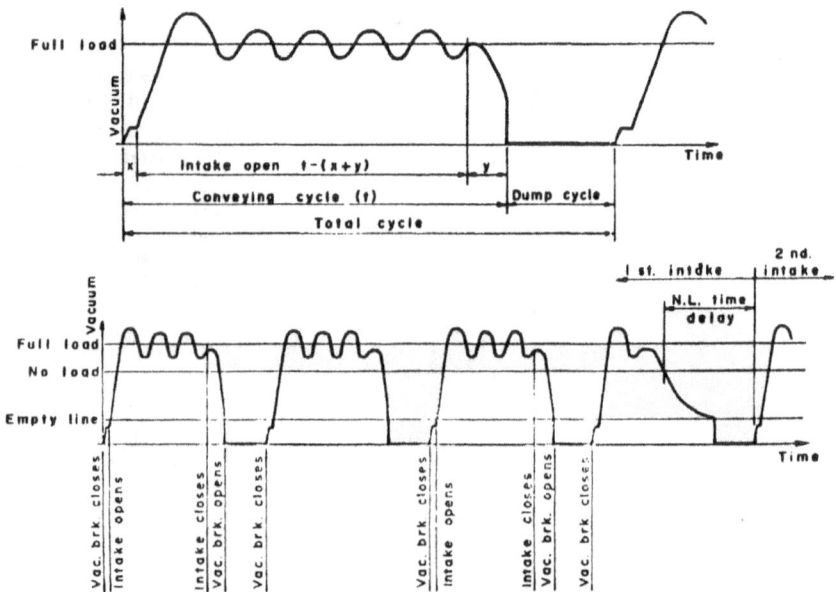

Figure 54

Total cycle, seconds	90	75	75	60	60
Conveying cycle, seconds	60	60	60	50	50
Dump cycle, seconds	30	15	15	10	10
X, seconds	2	2	2	2	2
Y, seconds	3	3	7	3	7
Cycle efficiency %	61	73.3	68	75	68

Cycle efficiency can be improved by using a shorter dumping cycle. The designer should be very careful when doing this. If the separator does not empty completely each time, it will eventually overload. When handling ash that is fluidizable, the dumping cycle can be cut to ten seconds, but only if the separator is externally fluidized.

Power or vacuum plus an air flow for system operation can be produced by a steam exhauster, water exhauster or by a negative pressure blower. To create the cycling effect, both the materials handling valve and the air flow are cycled constantly. A vacuum breaker alternately opened and closed to the atmosphere is used to cycle the air flow when the system operation is by a blower or by a water exhauster. The vacuum breaker and the materials handling valve are controlled electrically by a P.C. When a steam exhauster is used, the air flow is controlled by cycling the steam flow to the exhauster. Again, the cycling is controlled by a P.C.

Full load is defined as the system design vacuum, or vacuum high. When the system is energized, the water valve opens and the vacuum breaker valve closes. The vacuum level will rise to the full load, or vacuum high level. At this point, the controls will initiate the first materials handling valve, along with line segregating valves to open. As material is conveyed from this hopper, the vacuum level will fluctuate as shown in Figure 52. This will continue this way until the level of ash nears the bottom of the hopper. Then the vacuum level will begin to drop until it reaches a no load, or vacuum low point. The vacuum low setting should not be set so low that the system seems to be struggling to get to the set point. Since the hopper will not have any ash in it, the vacuum producer is just pulling flue gases from the hopper, and that is detrimental. Flue gases pulled through the vacuum producer mixing with the water could cause a chemical reaction that could be harmful for the equipment, as well as the disposal area.

The vacuum low switch is set to transfer from one hopper to the next. This signal closes the material handling valve. The vacuum will rise again until it reaches vacuum high and the next material handling valve opens. This procedure continues until all the hoppers in that row have been empted. The system will then step to the next row of hoppers closer to the vacuum producer. This sequence will continue until all of the hoppers have been empted. There normally is a purge time at the end of the operation to be certain that all of the ash has been conveyed from the lines.

The intermittent operating system and the continuous operating system with separating equipment consist of a primary separator, and sometimes a secondary separator. Steam exhausters, hydraulic vacuum producers, or a mechanical producer are the motive power. Maximum expected efficiency of this basic group of equipment is eighty-five to ninety-eight percent. When a mechanical vacuum producer is used, a bag filter is needed to collect to a higher percentage and also for protection of the mechanical vacuum producer. On some occasions, air washers or scrubbers have been used, but caution is to be exercised. Depending on the chemical analysis of the ash, or if flue gasses are pulled through the system, the combination could cause corrosion problems in the mechanical vacuum producer.

When bag filters are used, they must be designed for the maximum vacuum in the system.

When the air washer or scrubber is used, as the name implies, it uses water as a scrubbing agent.

The air steam is dry as it enters the washer but saturated as it leaves. The pipe between the washer and the mechanical vacuum producer is wet due to the condensing of the water vapor. It is, therefore, subject to pluggage from minute-sized particles that bypass the water.

CONTINUOUS OPERATING SYSTEM – WITH SEPARATING EQUIPMENT

A continuous operating system is used when the desired capacity cannot be obtained using an intermittent system. It is also used when the ash must be temporarily stored in a silo or day bin. The term continuous refers to the material being conveyed and collected continuously even though it is discharged on an intermittent basis.

The collecting and separating equipment must be of an airlock design. There is an upper and lower dump gate with a chamber between them of sufficient size to transfer the ash from the conveying system to the silo or day bin. There is an equalizer valve connected to the chamber between the two dump gates and the conveying line between the primary collector and the secondary collector. T here is another equalizer valve connected to the chamber between the two dump gates and the chamber below the lower dump gate.

The dump gates and equalizer valves are electrically controlled in the proper sequence to insure continuous material collection with intermittent discharge. See Figure 55.

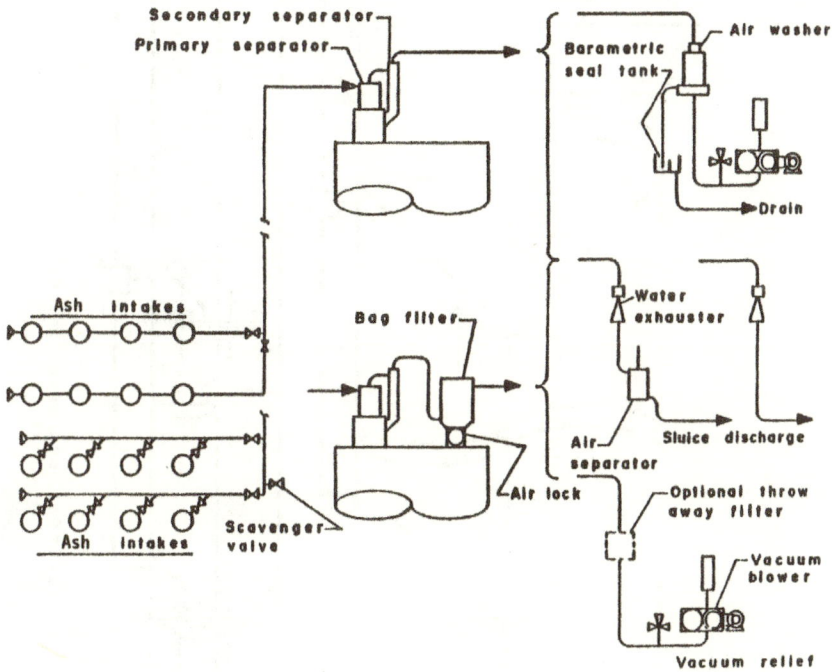

Figure 55

The basic operation of the continuous collector is fairly standard, but there are almost as many design variations as there are manufacturers.

The assembly usually includes a primary and a secondary collector as two separate structures. See Figures 56 and 57,

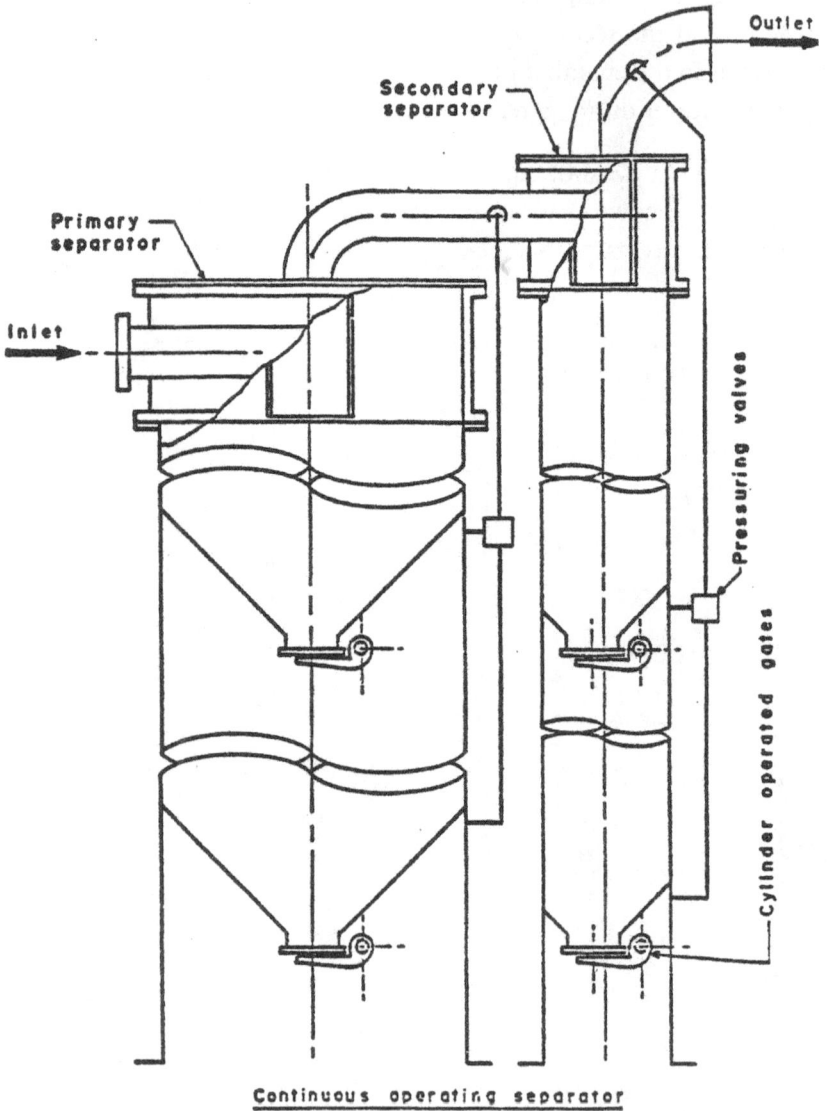

Figure 56

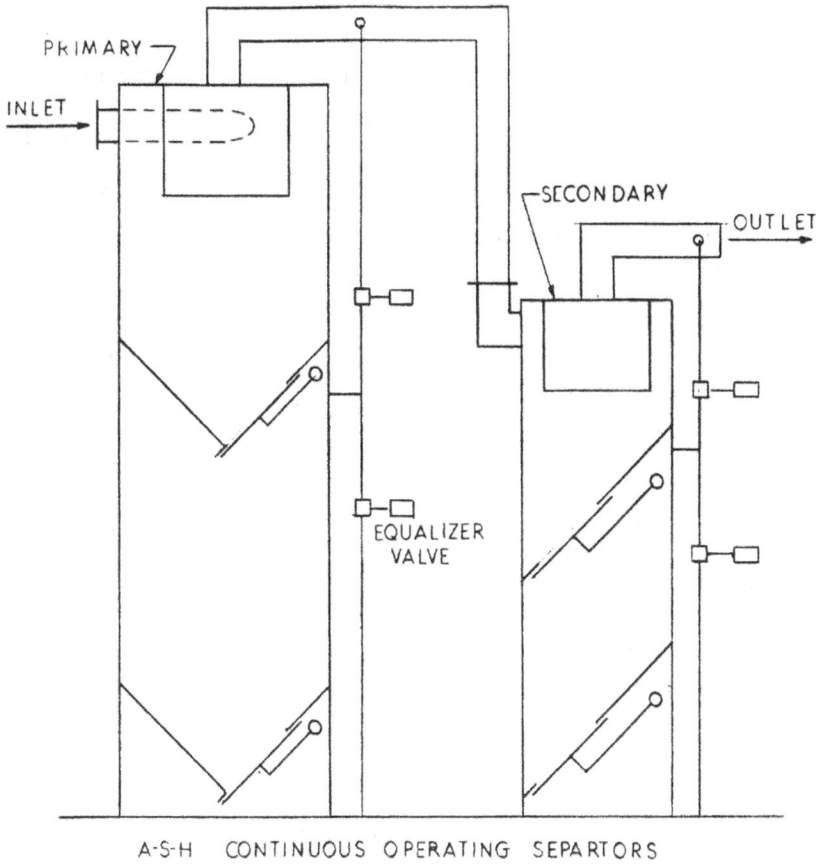

PRIMARY

INLET

SECONDARY

OUTLET

EQUALIZER
VALVE

A·S·H   CONTINUOUS OPERATING SEPARTORS

Figure 57

The continuous operating system may be used to convey any type of material as long as the correct design of collecting equipment is used.

Several different types of vacuum producers can be used: mechanical vacuum pumps, hydraulic vacuum producers, and steam exhausters. Steam exhausters, when used in the past, were limited to small industrial systems. Hydraulic vacuum producers are also being phased out due to water pollution laws. Mechanical vacuum producers are the most efficient, with either dry or liquid ring designs, using less horsepower.

The only control variation between the intermittent and continuous operation system with collection units is that of the feeding devices. In the intermittent operating system, the intake valves are continuously cycled to allow the dumping of a single compartment collector. In the continuous operating system, the intake calves are not cycled. The use of a double compartment collector allows for continuous material flow.

CHAPTER 12

POSITIVE PRESSURE PNEUMATIC CONVEYING SYSTEMS

A positive pressure conveying system has a variety of uses. There are many different manufacturers, and each has its own designs and variations. It can be a small or large system with many variations.

An air lock must be used to feed the material into the conveying line. The system may be either dilute phase or dense phase, and each type will be discussed separately.

A positive pressure conveying system operates at pressure levels above atmospheric. Although there is no maximum limit, 125 PSIG is about the limit most manufacturers use. Most dilute phase systems operate at pressures of 15 PSIG or below. This limit is arbitrarily set by the manufacturers to avoid the cost of pressure coding the vessels and to stay in the range of reasonably priced pressure blowers.

Some of the advantages of a pressure conveying system are: higher capacity and distances, the blower operates in clean air, blowers are positive displacement units, and the system may discharge to multiple points, such as silos, day bins, or a process, without any collection equipment.

The receiving vessel must be adequately vented, usually with a jet pulse type bag filter.

DILUTE PHASE FEEDERS

Dilute phase feeders generally have a single compartment with inlet and outlet gates. See Figures 58 and 59. They can be charged with material or they can discharge material, but they cannot do both at the same time. Therefore, from a single feed point, the flow is a batch operation. Thus, to obtain a continuous flow of material into a conveying line, two or more feeders must be operated on off setting cycles. The use of rotary feeders for fly ash has been discouraged due to the abrasiveness of the material, as well as the leakage past the rotor and the shoe of the feeder. Still, there are some manufactures that promote their use but normally at very low operating pressures and handling rates.

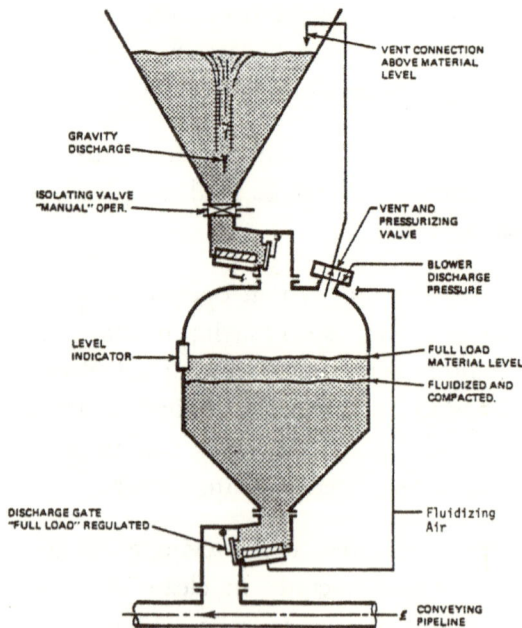

Figure 58

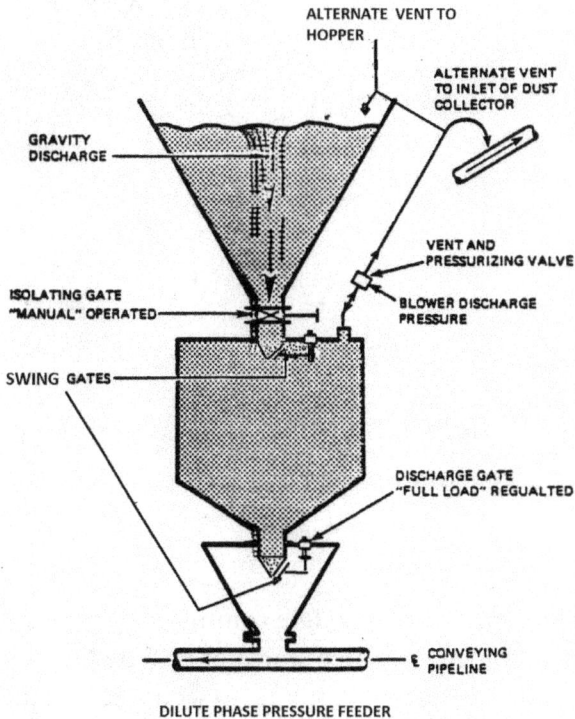

ALTERNATE VENT TO HOPPER

ALTERNATE VENT TO INLET OF DUST COLLECTOR

GRAVITY DISCHARGE

VENT AND PRESSURIZING VALVE

ISOLATING GATE "MANUAL" OPERATED

BLOWER DISCHARGE PRESSURE

SWING GATES

DISCHARGE GATE "FULL LOAD" REGUALTED

CONVEYING PIPELINE

DILUTE PHASE PRESSURE FEEDER

Figure 59

Systems that are used primarily in power plants, fly ash is collected in hoppers at the economizers, air heaters, precipitators, or bag house hoppers. The various parts of the boilers have groups of hoppers, and each group is isolated into small conveying system segments. In a segment, one hopper is conveying material to an air lock, while another is feeding material into the conveying line. Other air locks in that branch line are being vented waiting for their turn in the cycle to convey material. See Figure 60.

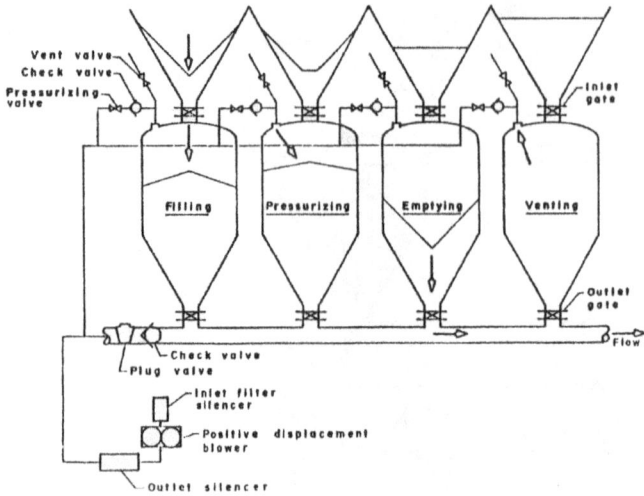

Figure 60

The hoppers in any given branch line continues to sequence in the cycle until all the material that has been stored or collected in the hoppers has been conveyed, or the hoppers continue to sequence by a predetermined number of cycles. See Figure 61, 62 and 63.

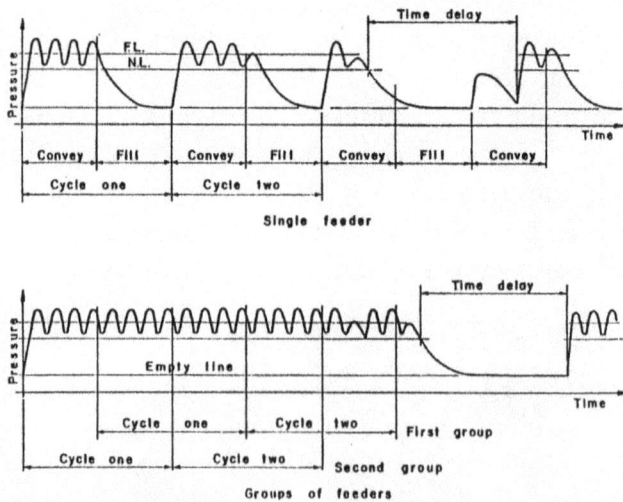

Figure 61

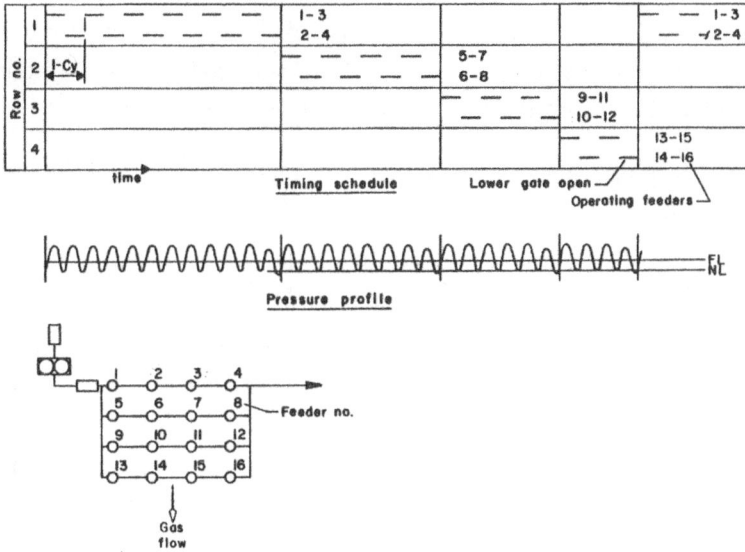

Figure 62

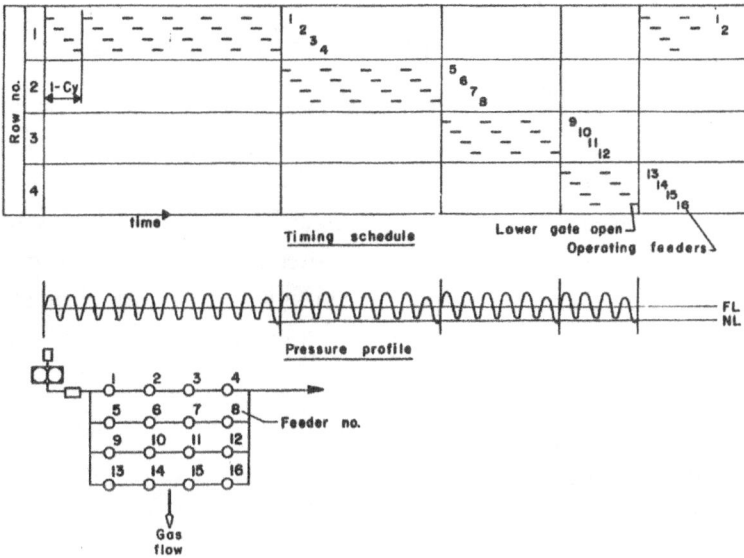

Figure 63

For a feeder to receive material it must be at a pressure equal to or less than the hopper in which it is connected. This is accomplished by a feeder vent system which allows the air in the feeder compartment and the displaced air from the incoming material to be vented to:

1. A point above the maximum material level in the hopper directly above the air lock.

2. Through a filter that connects the equalizer valve to the Air Lock Valve, then to atmosphere.

3. The inlet duct of the precipitator or bag house through a connecting pipe. The pressure and air flow within the vent pipe is induced by an inline fan and controlled by an inlet control valve to maintain the vent line pressure equal to or slightly above that of the feeding hopper. All of the vented air and its entrained dust must re-pass through the precipitator or bag house.

The system designer should familiarize himself with the above venting alternatives. Customer preference and/or other factors may dictate one verses another.

For an Air Lock to discharge material, its upper chamber must be at a slightly higher pressure than that of the conveying line. To accomplish this, some of the air from the blower discharge is connected to each air lock through an equalizer valve. This equalizer valve is opened based on the air lock control cycle. To insure the pressure entering the top of the air lock is higher than that of the conveying line, it is normal practice to use an orifice or plug valve between the blower and the first air lock. See Figure 60.

Theoretically, there is no maximum number of air locks that can be connected to any one branch line; however, most system suppliers limit the number to eight.

There are two pressure switches located at the discharge of the blower. The one switch is a high pressure switch or full load. The other is clear line pressure. The full load switch is set at the point where, if more material were introduced into the line and the pressure were to ruse, you run the risk of a plugged line, so all the bottom gates of the Air Lock Valves close and remain closed until the operating pressure drops below clear line pressure.

The clear line switch is used to advance cycle of air locks to the next air lock, or to the next branch line.

Hopper aeration is normally provided to aid the flow of material from the hopper into the upper chamber of the Air Lock Valves. The timing of the Air Lock Valves is to be set to empty the chamber between the upper and lower gates each time. If the chamber of the air lock is not completely emptied after each cycle, the material level within the air lock will become higher with each filling cycle. In some air lock designs, such as that shown in Figures 58, high level will interfere with the operation of the upper gate.

Either extra time should be added to be assured the chamber is emptied, or a high level detector should be used to prevent the chamber from filling to full.

In other Air Lock designs, such as that shown in Figure 59, the gate can close against the flow of material. Still, caution should be exercised not to overfill because the vent line and/or equalizer valve could be plugged. This design generally has ports near the top and bottom of the chamber for a manual check on filling and emptying times.

There is virtually, no precipitator or bag house system in operation where the flow or entrained dust is evenly distributed across the collectors; therefore, the amount of material collected in the hoppers under these collectors varies greatly. When these hoppers are connected to air locks arranged in branch lines perpendicular to the air or gas flow, the amount of material that each air lock must handle also varies. The controls must compensate for this irregularity in overall system capacity.

The number of air locks required and the location is generally set by the location of the collection hoppers. The conveying branch lines should be perpendicular to the direction of the gas or air flow. This is to equalize as much as possible the material distribution across the branch line. When conveying from a precipitator, the first row of hoppers will collect most of the material, generally about 80 percent. The hoppers of each succeeding branch line will collect approximately 80 percent of the remaining material.

It is general practice to set the capacity of the system equal to two times that of the maximum production, plus an allowance for down time.

DENSE PHASE FEEDERS

There is great confusion in Dilute Phase Conveying versus Dense Phase Conveying, both in literature, and industry as to the proper usage of these two terms, however, the most important aspect of dilute versus dense phase is the materials behavior during pneumatic conveyance.

Where dilute phase conveying is when the material being conveyed is fully suspended in horizontal pipelines, and generally below ten to one (10:1) phase density, dense phase conveying has a higher phase density, and can have numerous flow patterns. See Figure 64. Dilute phase conveying may be referred to as lean phase or suspension flow. Along with Phase Density, most references will establish ranges of velocity, as well as the phase density. Depending on the reference, there is a wide range of velocities for common materials.

Phase density is the ratio of the mass flow of the solids being conveyed to the mass flow rate of the Conveying gas. Phase density is used to describe the nature of the gas-solids flow in a pipeline. Other terms used to describe this relationship include Solids Loading Ratio and Mass Flow Ratio (pounds of material/pounds of gas) Fly ash for an example, as a simple term for a material has a very wide range of conveying velocity requirements due to many factors. Chemical analysis, particle size distribution, particle shape, and density are among some of these factors.

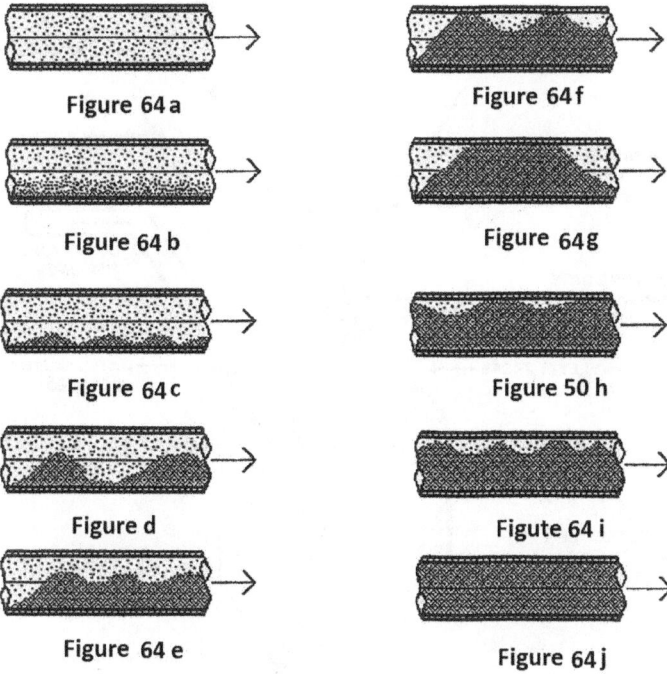

Figure 64 a

Figure 64 f

Figure 64 b

Figure 64 g

Figure 64 c

Figure 50 h

Figure d

Figute 64 i

Figure 64 e

Figure 64 j

Figure 64

There are many manufacturers of dense phase feeders like dilute phase, and each has its concepts of how fly ash is to be handled. Most suppliers agree that fly ash should be fluidized before it can be conveyed. The vast majority of dense phase feeders employ a fluidizing element in the extreme bottom of the feeder. Once the fly ash has been fluidized, and the vessel pressurized, the fly ash starts moving down the conveying line to the discharge point. In all cases, there is a discharge valve on the feeder. Discharge valves are required when there are several feeders connected to one discharge line. Figure 65 shows a typical dense phase feeder arrangement.

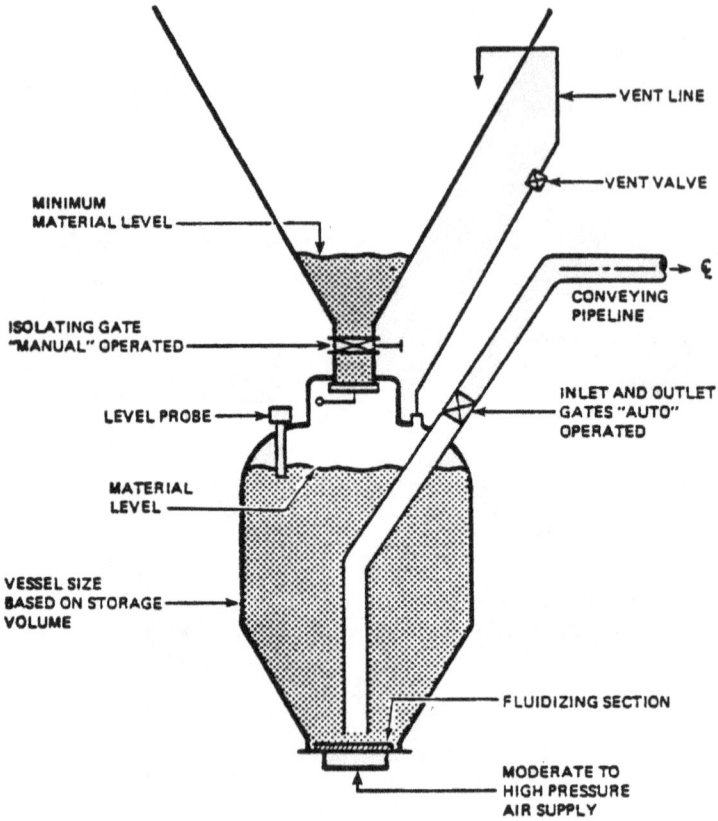

DENSE PHASE PRESSURE FEEDER

Figure 65

Dense phase system operation is characterized by the conveyance of material below the saltation velocity, again reference Figure 64. The material is fluidized in the Transporter Tank and fed into the pipeline through the internal pick-up tube. Some of the conveying air is introduced as fluidization air and the ratio of this air to the total flow is known as TTAR (Transporter Tank Air Ratio). This ratio is pre-determined later in this procedure. This balance (sometimes very critical and sensitive) must be determined prior to selection and design of the critical flow nozzles. This ratio controls not only the air flow split, but the feed rate of material into the conveying line. The basic function is to move material from a zone of lower pressure such as a hopper, bin, or silo into a zone of higher pressure such as the conveying line. This transfer is generally accomplished in a batch mode of operation. The amount of material which can be conveyed in any given pipeline configuration is dependent upon the pipeline geometry, characteristics of the material being transported, the available gas flow, and pressure.

Pipeline diameters are generally smaller, conveying velocities lower, the pounds of material per pound of air much higher, and the operating pressure higher. It is only possible to discharge one feeder at a time. Since the operating pressure is that required to push a fluidized slug of fly ash a given length through a constant diameter pipe, it remains fairly constant until the slug starts to emerge from the end of the conveying line. As the slug emerges, the pressure tends to drop and the velocity increases. As the slug leaves the pipeline, a considerable surge of airflow (SCFM) to the receiving tank occurs. The pressure switch on the dense phase feeder should be set to sense the drop in pressure and sequence to keep the airflow to a minimum. This drop in pressure occurs when the feeder and pipeline are empty. T he vent system on the receiving tank is considerably smaller than that required for a dilute phase system due to the lower airflow (SCFM) requirement.

After each feeder has been emptied of fly ash, it must be vented to hopper pressure or lower before it can be refilled with fly ash. Some manufacturers pressurize the feeder by an air connection above the material and others pressurize and fluidize through the fluidizing membrane.

Dense phase systems are not subject to the same limitations upon vacuum systems, but are not as restricted as the typical "dilute phase pressure systems:. Dense phase conveying systems are generally restricted to fluidizable material and the materials ability to retain air determines the conveying distance the material can be successfully transported.

More critical to the system capacity than transport line capability is the rate at which material will flow from the Transporter Tank. This rate is dependent upon many factors including tank geometry, material chemistry, and the physical properties of the material. The physical properties include the materials ability to fluidize. Particle distribution is also a critical characteristic and the material should be 90% less than 300 microns in size. The relationship of these variables is very complex; therefore, each specific design requires a detailed analysis by a experienced personnel.

Application of the design standards is based on flow which is below the saltation velocity. Dilute phase conveying is above the saltation velocity. Dense phase usually has a Solids Loading Ratio (SLR) above 20:1 whereas dilute phase is generally below 15:1 and most times 10:1.

Dense Phase system pressure ranges generally require an air compressor. The compression is sized to provide a minimum pressure at least 20% higher than the required conveying system pressure. This pressure drop is a requirement of the critical flow nozzles which controls the air flow to the Transporter, as well as the conveying line with an additional 20% for safety margin. Site elevations above 2000 feet require special consideration due to the reduction in atmospheric pressure and associated reduction in air density.

Selection of the appropriate minimum pickup velocity is critical to proper system design. Experience indicates that the following pickup velocities for the materials as noted:

Pulverized Coal Fired Boilers

Ash with a combined CaO & MgO less than 4% 1200 feet/minute
Ash with a combined CaO & MgO more than 8% 1600 feet/minute

NOTE: For any specific application, the proper pickup velocity must be established by having a sample of the material tested in a qualifying laboratory.

In multiple line size systems, the velocity at each increase in line size must be at least 90% of the selected minimum pickup velocity.

Precipitator hopper vibrators are not to be used or recommended for use with dense phase systems. Vibrators will only compact the ash above the Transporter inlet valve even if attempts are made to time the vibrator operation to periods when the Transporter inlet valve is open. Fluidization of the hoppers with aeration air is required to establish flow of ash into the Transporter when the top inlet valve is open and should always be furnished as part of the dense phase system design.

Intermediate high hopper level probes may be required to provide advance warning of Transporter malfunction. The intermediate hopper high level probe should be located at a level which provides sufficient time to correct any difficulty before the hopper dust level rises to the point that it will impair the precipitator or bag house operation.

The TTAR is determined when testing the material in the Laboratory. This ratio is the ratio of the air entering the Transporter Tank to fluidize the material versus the supplementary air. This balance (sometimes very critical and sensitive) must be determined prior to selection and design of the critical flow nozzles. This ratio controls not only the air flow split, but the feed rate of material into the conveying line.

Proper system design requires an interpretation of the customer's specification in association with a proper analysis of the system operating conditions. Dense phase system capacity is most affected by the Transporter operation and the fact that the Transporters are volumetric devices operating in a batch mode. The system can only remove material which is available for transport. A high efficiency operation requires sufficient material in the feed hoppers to fully fill the Transporter during its cycle. This is seldom the case for lightly loaded rows located on the back end of the precipitator or bag house.

Additionally, the flow rate through any Transporter is dependent upon the flow characteristics of the material and the Transporter outlet size.

For precipitators, the system controls are designed with a row count feature which provides a greater number of cycles for the heavy precipitator rows versus the lightly loaded rows. Minimum row count for any row is two. This control scheme provides an operational margin of safety to insure complete ash removal from the entire precipitator, but adversely affects the system efficiency. This must be understood and accounted for in any performance testing to avoid negative impact of such operation on the performance test.

TRANSPORTER DESIGN AND SIZING CRITERIA

The pressure required for conveyance is the system design pressure and as stated above, the nozzles require a pressure 20% higher than this pressure. The air supply is sized for a pressure 20% higher than the nozzle pressure as a safety factor (i.e. the air supply is sized for the design pressure plus 44%).

Since the discharge is out the top, a fluidizable material is necessary. The Transporter Tank is designed with a fluidizing media at the bottom to fluidize the material, and the conveying is normally higher than 15 PSIG which dictates that the Transporter Tank be a coded pressure vessel. With top discharge, there will be a pickup tube which will exit near the top of the Transporter and be very close to the fluidizing media in the pan at the bottom of the Transporter. This pickup tube is adjustable at the bottom to allow for different gaps between the bottom of the pickup tube and the fluidizing media.

The Transporter is normally cycled continuously. For use with precipitators or bag houses, the Transporter will sequence similar to Air Locks. The main component of the system is the Transporter Tank. The material inlet connection is located on top of the Transporter along with two smaller openings. These smaller openings are for the vent connection and gauge connection (if required).

The vent line piping must be as vertical as possible with no horizontal sections.

Since the velocities are much lower than for dilute phase conveying, standard steel pipe can be used. T he bends are to be rolled, and are to have a bend ratio of the following:

"Centerline bend diameter / pipe diameter" = 15/1.

"Radius of the centerline of the bend / pipe diameter" = 7.5/1, or the bend radius is 7.5 times the diameter of the inside pipe diameter.

AIR REQUIREMENTS

The air supply must consist of clean dry air. For most dense phase systems, the required pressure will be above that achievable with positive displacement blowers; therefore, compressors are used.

The air supply subsystem consists of: Compressor, after cooler, water/ oil separator, dryer, air regulator with a pilot regulator and a cylinder operated butterfly valve to control the air flow. See Figure 66.

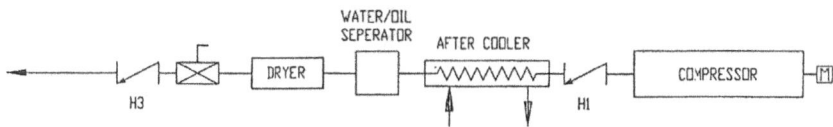

Figure 66

Critical flow nozzles (choke flow) control the air flow for each Transporter and supplementary air source. These nozzles require an upstream pressure at least 20% higher than the required design conveying pressure.

The pressure required for conveyance is the system design pressure plus a pressure 20% higher than the nozzle pressure as a safety factor (i.e. the air supply is sized for the design pressure plus 44%).

CONTROL ITEMS

A schematic of the controls is shown in Figure 67 for a single unit system. The Transporter is normally cycled continuously with only one dumping at a time. For precipitators, the system controls are designed with a row count feature which provides for the capability of addressing the heavy precipitator rows a greater number of cycles versus the lightly loaded rows. Minimum row count for any row is two. This control scheme provides an operational margin of safety to insure complete ash removal from the entire precipitator. This adversely affects the system efficiency which must be understood and accounted for prior to performance testing to avoid any negative impact on the performance test.

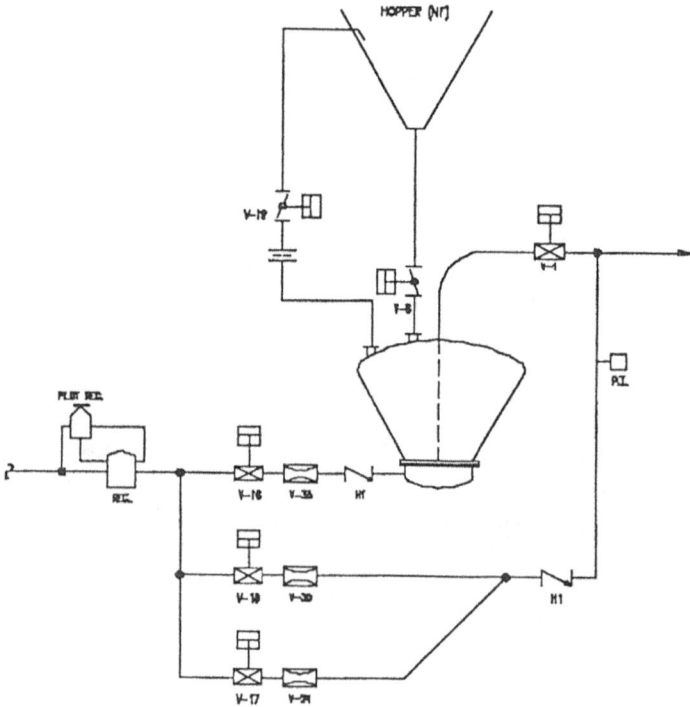

Figure 67

The control items are listed below:

1. Inlet valve (cylinder operated) (V-8)

2. Discharge valve (cylinder operated) (V-1)

3. Fluidizing air supply valve (cylinder operated) (V-16)

4. Supplementary air supply valve (cylinder operated) (V-17)

5. Transporter Tank by-pass air supply valve (cylinder operated) (V-18)

6. Vent valve (cylinder operated) (V-17)

7. Fluidizing air balance valve for fluidizing in the lower cone (manual valve) (H-7)

8. Fluidizing air balance valve for fluidizing at the back of the Feed pan (manual valve) (H-6)

9. Critical flow nozzles for the fluidization air tank by-pass air, and supplementary air (V-33 & V-24)

10. Pressure sensors for the air supply upstream of the critical flow nozzles. (P-5)

11. Main air supply pressure regulator incorporating a feedback pilot

12. One way valves (check valves) in each air supply connection (H-4, H-5 & H-6)

The critical flow nozzles require twenty-four (24) pipe diameters of straight pipe between the control regulator and the nozzles.

A single Transporter operates in a batch mode. The controls are set to operate continuously even though conveying is a batch operation.

When the system is initiated, the supplementary air valve (V-17) and the tank by-pass air valve (V-18) are opened. The conveying cycle begins by opening the vent valve (V-19) on the Transporter. After a set time (to ensure pressure equalization between the Transporter Tank and the hopper approximately 5 seconds), the inlet valve (V-8) opens. Material fills the Transporter and the inlet valve (V-8) closes based on a timer setting. At this time the fluidization air supply valve (V-16) opens. This fluidizes the material in the tank and also pressurizes the tank. After 5 seconds the discharge valve (V-1) opens and the tank by-pass air valve (V-18) simultaneously closes. The material is then conveyed through the pipeline.

As the material level in the Transporter Tank lowers, the feed rate decreases and the conveying line pressure lowers. As the Transporter empties, the operating pressure continues to fall until a predetermined conveying pressure {Convey Lo} is reached. This signals the discharge valve (V-1) to close and simultaneously opens the tank by-pass air valve (V-18). T he vent valve (V-19) opens and the cycle repeats until the conveying process is terminated. The system then continues to purge the line for a specified period after which the supplementary air valve (V-17) and tank by-pass air valve (V-18) close.

Multiple unit operation is continuous and operates in a manner similar to a single unit (see Figures 67, 68 and 69).

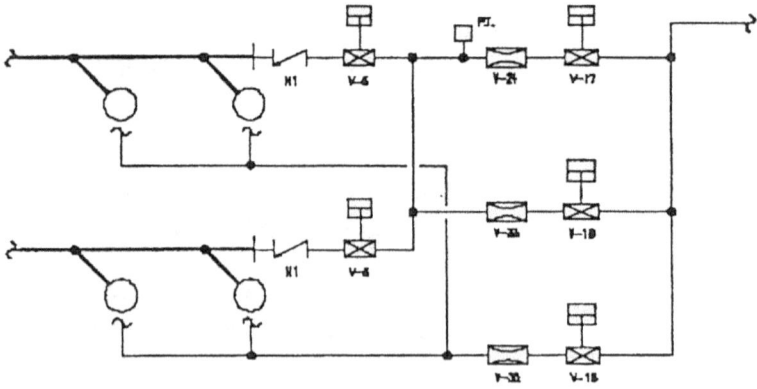

Figure 68

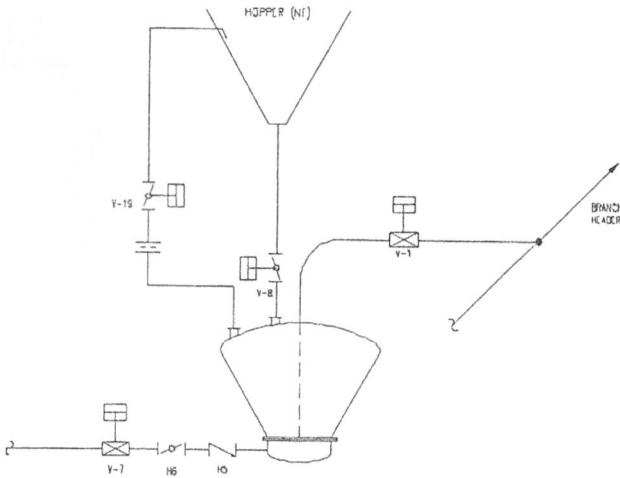

Figure 69

When the system is initiated, the supplementary air valve (V-17) and the tank by-pass air valve (V-18) are opened. The conveying cycle begins by opening the vent valve (V-19) on the first Transporter. After a set time (to ensure pressure equalization between the Transporter Tank and the hopper), the first inlet valve (V-8) on the tank opens. Material fills the Transporter based on a timer setting. The inlet valve (V-8) closes. At thus time, the fluidizing air supply valve (V-16) for the first Transporter Tank opens. This fluidizes the material in the tank and also pressurizes the tank. Once a predetermined tank pressure {Tank H-1} in the Transporter is reached, the discharge valve (V-1) opens and the tank by-pass air valve (V-18) simultaneously close. The material is then conveyed through the pipeline.

As the material level in the first Transporter Tank lowers, the feed rate decreases and the conveying pressure lowers to a predetermined value {Convey Med}. This signals the second Transporter Tank to commence the vent/fill cycle. After a predetermined time the inlet valve (V-8) closes on the second tank and the fluidizing air supply valve (V-16) on the second Transporter opens. Concurrently, the first Transporter empties and the operating pressure continues to fall until a second predetermined conveying pressure {Coney Lo} is reached. This signals the discharge valve (V-1) to close and simultaneously opens the tank by-pass air valve (V-18). The tank pressure in the second Transporter rises and after 5 seconds, the second discharge valve opens (V-1) and the tank by-pass air valve (V-18) simultaneously closes. This provides a nearly continuous feed of material into the conveying line.

This cycle proceeds sequentially through the branch until each Transporter in the row has been cycled at least once. At that time, the tank by-pass air valve (V-18) opens to maintain conveying air velocity while the branch is purged and the valves are aligned for the next branch. The conveying process continues on in this way.

The fluidization air supply valve (V-16) controls air flow to the Transporter. There is a time delay between the material inlet valve (V-8) closing and the fluidizing air supply valve (V-16) opening. This time delay must be kept to a minimum.

In both the fluidizing air line and the supplementary air line, there is a critical flow nozzle to control air flow. The sizing of the critical flow nozzle is based on the total conveying air requirement and the TTAR.

There is a one way valve in each of the air supply lines to prevent any material from flowing back into the air supply lines.

One material inlet valve (V-8) controls the flow of material into the material into the Transporter.

A vent valve (V-19) is required to vent the pressure and the displaced air from the Transporter Tank back into the hopper or bin prior to and during filling of the Transporter Tank. This valve remains open the full time the material inlet valve (V-8) is open.

Pressure sensors are used for the determining and recording the operation of the system. These sensors, with pressure settings in the control (PLC) system, are the means for controlling valve operation. In the controls, there is a pressure setting, higher than the operating pressure, to close the discharge valve and allow air to purge the system. Pressure settings are established by the Mechanical System Engineer. The high pressure setting is set at 20% higher than the system design pressure.

A discharge valve (V-1) is used on each Transporter and opens at a predetermined time. The discharge valve (V-1) closes at a predetermined descending pressure in the conveying line.

Pressure settings are predetermined based on the system configuration and design requirements. The pressure sensor controls the opening of the discharge valve, fluidization valve, supplementary air supply valve, and (at the end of the cycle) opening of the vent valve. All of these pressure settings are established by the Mechanical System Engineer.

TRANSPORT SYSTEM PIPING CONFIGURATION

No more than two rows of hoppers (running parallel or perpendicular to the gas flow) may be "looped" to form a branch line. This arrangement may be desirable to simplify design, operation and control of the system. Generally, the conveying system branch lines should be run perpendicular to the gas flow to achieve maximum system capacity and efficiency as this group's hopper having the same loading on the same branch line. In cases where it is desirable to run the branch lines parallel to the gas flow, a "looped" arrangement must be used.

Ash transport piping furnished is usually steel pipe with bends made in accordance stated in the "Transporter Design and Sizing Criteria" section. Pipe and fittings can be plain end and welded. The piping system must be rigidly clamped and supported. A manual isolation devise is usually furnished between the dust hopper discharge and the Transporter inlet as isolation for maintenance if necessary. Conveying lines are to be horizontal or vertical only. Sloping lines are not permitted. The number of bends in ant line should be minimized. At least 10 pipe diameters of straight pipe downstream of any pickup point before any directional change is made, is required to provide stabilization of material flow. Any valving, crossovers, and tie-ins between lines should be arranged to minimize or prevent buildup in the system "dead leg".

Multiple discharge lines requiring the use of valves in a crossover application also present severe application problems. These valves must be capable of bidirectional seating as pressure may be applied to both sides of the valve depending upon the flow path selected. Resilient seated valves may be considered for this application provided the valve seat material is adequate for the expected material and air temperature. Crossover valve alignment should be set prior to system operation initiation and cycling of the valves during the transporting of ash or as a part of the normal collection sequence should be avoided.

CHAPTER 13

Vacuum - Pressure Pneumatic Conveying Systems

As previously mentioned, it is sometimes economical to consider a combination of vacuum and pressure conveying of fly ash where distances rule out the use of vacuum alone. This type of system permits the use of vacuum to remove ash from hoppers at a high rate to a collecting point nearby where the ash is continuously transferred to a pressurized conveyor. Thus, both types of systems are utilized to their best advantage. The vacuum system, with its simplified controls can remove fly ash at an optimum tate within the power plant itself. The pressure system, reduced to one transfer point with a minimum of controls, can then deliver the collected fly ash to any terminal point on the power plant property up to several thousand feet.

Figure 70 shows portions of these systems adjacent to the transfer points. This arrangement is for fly ash from typical precipitators, or bag house collection. Figure 71 shows the same portions for fly ash that has a higher temperature than that of the typical precipitator, or bag house.

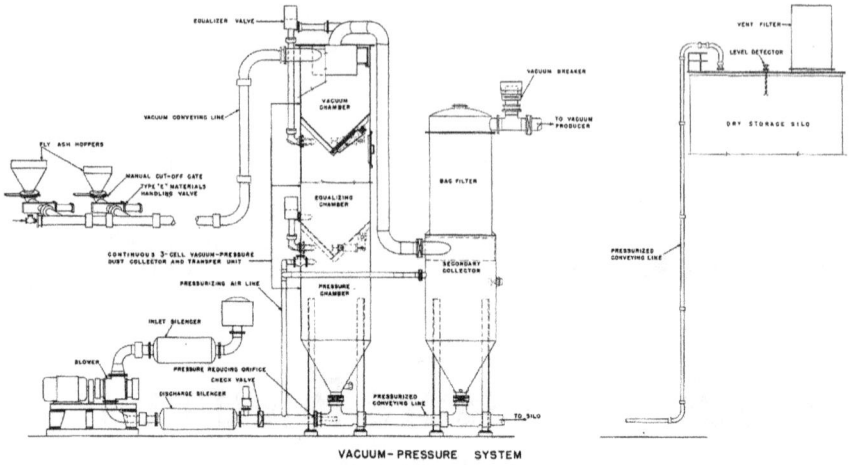

Figure 70

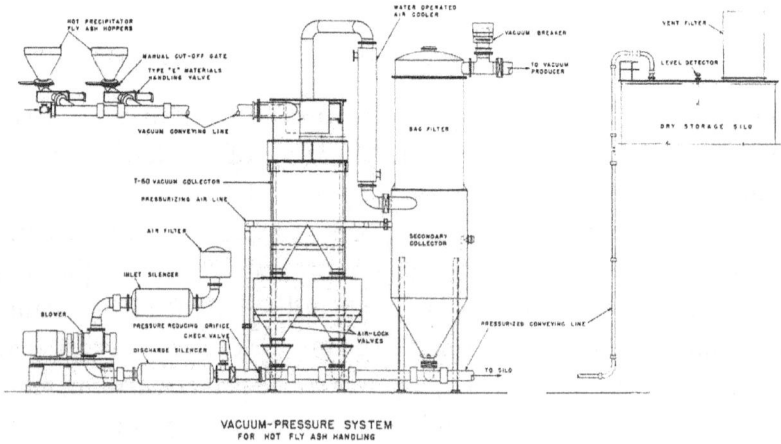

Figure 71

Fly ash is removed from the hoppers through air-electric operated Materials Handling Valves as described for "Vacuum Systems". Any of the vacuum producers previously described may be used to create the vacuum required for this part of the system. Fly ash is dropped out of the air stream in a series of collectors. The primary collector is provided with three chambers or cells. By means of interlocked gates and equalizing valves fly ash is passed, without interruption, from the vacuum system, through an equaling chamber, into a pressure chamber from which it is blown continuously onto the conveying pipe.

The vacuum line from the primary collector discharges into the inlet of a secondary collector. The lower portion of the secondary collector is a cyclone which drops a considerable amount of fly ash carryover into a storage chamber below it. Any particulate which remains in the air stream is trapped by a self-cleaning bag filter above the cyclone collector. The secondary collector is dumped into the conveying pipe only as often as experience indicates necessary. In some cases this is required only at the end of a dust handling cycle.

A blower provides the air flow and pressure required to convey the fly ash at the required rate from the three-cell transfer unit and secondary collector. By properly sizing nozzles and orifices the pressure in the conveying line is kept about one PSIG less than in the pressure chamber of the transfer unit. Thus, fly ash is forced to flow from the chamber into the conveying pipe.

Where fly ash is conveyed from hot precipitators some modification to the above is required although the basic principle remains the same. The primary cyclone collector discharges through a pair of Air-lock valves equipped with heat resistant parts to withstand the elevated temperature of the fly ash. An air cooler is installed ahead of the secondary collector to protect the bag filter by reducing air temperature to a level the bags can tolerate.

The vacuum-pressure system for the least complex controls of any long distance pneumatic conveying system by combining the simplicity of vacuum system controls with those of a two point or three point pressure system. This contrasts with the complexity of controls required by an all pressure system using a multiplicity of Air-locks to achieve the same conveying rates.

CHAPTER 14

Pneumatic System Design Calculations

In this segment we will discuss one method for the calculations of a pneumatic conveying system. First of all it is necessary to use an airflow equation that is compatible to the lengths of the line, and pressures normally encountered. The design of a pneumatic conveying system is almost as much of an art as it is an engineering function. Investigators have studied the subject for years, and many papers have been written on the subject. To date, there is no single accepted method of calculation nor even one universally accepted explanation of the basic theory. There are many solutions in use, and many explanations of the mechanics of the conveying process. Very few of the calculating methods used give the same identical results. This book will not attempt to explain any theory, but will present a method of calculation for the dilute phase system.

One point generally agreed upon is that the total system losses are composed of two separate and distinct parts: first, the losses due to air flow alone and second, the losses associated with the addition of the material. For lightly loaded systems, such as vacuum cleaning and venting in the dilute phase realm, the addition of the material has little effect on the total losses. But, as the quantity of material is increased in relation to the air quantity, the added losses can be significant. In a dense phase system, the losses due to the material are the total system losses for all practical purposes.

The first consideration in a dilute phase system is the required air velocity to move the material through the pipeline. Some of the required velocities are listed in Table 4 in (Chapter 10), for various materials. It is suggested that anyone designing a system use them only as a guide, and obtain suitable values by a test program. This can be done very simply by calibrating air flow through a glass pipe. If operated at less than atmospheric pressure, vacuum, the observer can add material to the air flow without the use of an air lock device. By visual inspection alone, the observer can tell how the material reacts to various velocities. The test setup for determining minimum pickup velocities does not have to represent a particular system configuration, but it does have to be conducted with the actual material to be conveyed. By including both a horizontal and vertical test section, the observer can obtain the values of both. The velocity requirements in a vertical pipe will always be less than in a horizontal pipe. By use of glass, the observer can also determine the reaction to material flow with elbows of various radiuses and types. The velocity that is of concern is not the one that barely moves the material, but one capable of reentraining material from the bottom of a pipeline. It should be well above saltation velocity to prevent slugging or drifting in the system.

For air alone, the pressure drop is proportional to the square of the velocity. To this must be added a factor for the material loading. Therefore, the velocity becomes a very important consideration in that the total system losses are dependent on its value. The lower its value, the lower the total system losses will be. The designer is cautioned against setting the initial velocity too low. To do this would result in a plugged system or one that is not capable of conveying the material at the capacity and at the designed pressure. The only field answer to this condition is to reduce capacity and pressure or to install a larger power source to give greater air flow and the resultant higher power requirements. Should it develop in the field that the chosen velocity is higher than that required to transport the material safely, there are several alternatives that can be used:

1. The velocity can be lowered by slowing down the blower, reducing the water gpm (liter/s) or pressure if a water exhauster is used, or by introducing an air leak in the system.

2. Operate the system at a higher loading by setting the full load control higher or by speeding up the device.

Either of these are better choices than having to replace the power source because the design velocities are too low.

A pressure drop or power loss is associated with the acceleration of the material as it enters the pipeline, by a directional change, and by elevating the material. About the only variable that the designer has any control over is the choice of the elbows and the type of pipe to use. With nonabrasive materials it is common practice to use elbows with long radii For abrasive materials it is much better to use long-radius elbows. First, the wear is concentrated in a short section and second, they eliminate slugging, which should be avoided at all cost. Its effect is much more pronounced in a positive pressure system than it is in a negative pressure system. Should slugging occur, the resulting pressure surges can be great enough to blow relief valves. More important, the rigid anchors must be heavy enough to hold the thrust loads. It is better to design the system to eliminate all slugging.

For non-abrasives, the use of steel pipe or even aluminum pipe can be justified. Steel pipe can also be used for abrasive materials if it is in the powdered form and can be considered to flow in the homogeneous range. Even though steel pipe can be used for both horizontal and vertical pipe runs, wear-resisting elbows must be used. As the flow direction changes, the particles come in direct contact with the inner walls, causing impact and a sliding action that can cause rapid wear. It is suggested that long-radius elbows of hard iron followed by a 3 foot section of hard iron pipe be used. This will contain most of the wear associated with the elbow plus the re-entrainment of the material. For coarser material and heavy solids where flow can be considered as heterogeneous, hard iron pipe or other abrasive pipe materials must be used. The surface roughness of the pipe has a considerable influence on the pressure losses of the system and should be taken into account in system design.

Air or gas flow through long pipelines can be calculated by using isothermal conditions. Pressure drops in conveying systems, of the type discussed in this book, are relatively large in relation to the inlet pressure.

Here Are Design Considerations

From the design point of view, there is no fundamental difference between suction and pressure systems, the choice generally being dependent on plant requirements and a consideration of economy. Design is basically a problem in determining the energy requirements. These can be expressed in pressure units, and from them, the size of the blower and required horsepower can be estimated.

Summation of the following five factors, in the consistent inertia of solids, to get them into motion:

1. Acceleration energy needed to overcome inertia of the solids, and get them into motion.

2. Energy required elevating the solids.

3. Energy, required to sustain solid materials, in a fluid stream, and to overcome the material's friction in the conveying pipe.

4. Energy losses associated with changes of direction at bends and elbows.

5. Fluid losses in the pipe, and at the terminals of the system. These are made up of pressure loss at the system's entrance, friction loss of the pure air stream in the pipe line, pressure drop across the collector at the end of the system, and losses from any auxiliary units which may be incorporated into the system. Fluid head and kinetic energy terms are generally neglected, unless the system is characterized by unusually large vertical lifts or fluid velocities.

Four of these design components are based on an important variable, velocity. Many references in the literature describe how to determine conveying velocity of solid materials. Theoretical velocity for conveying fine particles can be calculated with Stokes' law; Newton's law holds for large particles. Other investigators have suggested empirical formulas of more or less complexity.

All of these methods give only the critical or balancing velocities. So the designer usually assigns an additional empirical value to insure the sustained movement of the particles. Here, judgment comes into play; however, in table 4, in Chapter 10 offers a list of adequate conveying velocities. This table suggests velocity ranges from 1800 to 7000 ft/min., depending on the bulk density of the material, and probably in part on the particle size and configuration.

For the purposes of this book, and for nearly all conveying systems involving materials being conveyed from collection points from power plant systems, and with bulk densities ranging from 25 to 90 lb./cubic foot, a velocity of 4500 ft./min. will be found satisfactory. This may seem excessively high speed for some materials, however, the particles in an air stream rarely attain more than 80% of the theoretical air velocity, and in short systems, even less.

Now let's look at the design factors in greater detail. First we must find the specific volume of air at the inlet, then the specific volume of air at the discharge.. We establish a pipe diameter, and with the initial velocity, we can determine the discharge velocity, then the average which is used to calculate the system capacity.

The following symbols will be used:

M = lbs. / minute of material conveyed
V = velocity in ft. / minute
D = horizontal distance in feet
H = vertical distance in feet
P_1 = 14.6963 psia for vacuum systems – selected pressure psia for pressure systems

$P_2 =$ Selected pressure psia for vacuum systems – 14.6963 psia for pressure systems

$V_1 =$ Specific volume of air at dust inlet conditions

$V_2 =$ Specific volume of air at dust discharge conditions

$T =$ 1.2 for isentropic expression

For a typical vacuum system problem we will use the following system parameters:

30 TPH or 1000 lbs./ min., V= 1800 ft/min. minimum velocity D= 350 ft, H=100ft.,f=.7, 5 90° elbows, assume 6" dia. Pipe, 12" Hg vacuum, 250° temperature at inlet

First we want to find the specific volume at the dust inlet.

$$V_1 = \frac{1}{0.080728 \times P/14.6963 \times [491.6/(T_2 + 459.6)]}$$

$$= \frac{1}{0.080728 \times 14.6963/14.6063 \times [491.6/(250+459.6)]}$$

$V_1 = 17.88$ ft.^{3}/lb. at inlet

Now we want to find the specific volume at the discharge.

$P_1V_1^{1.2} = P_2V_2^{1.2}$ $P_2 = 12"$ Hg. $\times 0.4912 = 5.89$

$V_2 = \sqrt[1.2]{14.6963 \times 17.88^{1.2}/ 5.89}$

$V_2 = 38.28$ ft.^{3} / lb. at discharge

6" pipe A = 0.196 ft^2 .196 $\times$ 1800 = 353 CFM inlet

353 CFM / 17.38 = 19.74 Lbs/Min air $\times$ 38.28 ft^3/lb = 756 CFM at discharge

756 CFM / .196 ft^2 = 3857 ft/min Max

Min. Vel. 1800 ft/min

Max. Vel. 3854 ft/min

Avg. Vel. 2827 ft/min

Inertia = $MV^2/2g$ = M $(2829)^2$ / 231840 =   34.5 M

Vertical Lift = MH =   100 M

Horizontal Work = M D f = M 350 × .7 =   245 M

Fittings = MV^2 × π × f N /2g =   379 M

Total M   759 M

Air Work = $144(P_1V_1 - P_2V_2$ / Γ- 1) = 144 (14.6963 × 17.88 − 5.89 × 38.28/1.2-1) = 26,722 × 19.74 = 522,957

522,957 lbs/min / 759 M = 979 lbs/min or 29 TPH

For a typical pressure system problem we will use the following system parameters.

50 TPH or 1000 lbs./min., V = 2300 ft/min. minimum velocity D= 350 ft, H=100 ft, f=.7, 5 90° elbows, assume 8" dia. Pipe, 12 PSIG pressure, 250° temperature at inlet.

First we want to find the specific volume at the dust inlet.

$$V_1 = \frac{1}{0.080728 \times P/14.6963 \times [491.6 / (T_2 + 459.6)]}$$

$$= \frac{1}{0.080728 \times 26.6963/14.6063 \times [491.6/(250+459.6)]}$$

V_1= 9.84 ft³/lb at inlet

Now we want to find the specific volume at the discharge

$$P_1V_1^{1.2} = P_2V_2^{1.2} \quad P_2 = 12 \text{ PSI.}$$

$$V2 = \sqrt[1.2]{26.6963 \times 9.84^{1.2}/\ 12}$$

V2 = 34.66 ft.^{3} / lb. at discharge

8" pipe A = 0.349 ft^2 .349 × 2300 = 803 CFM inlet

803 CFM / 9.84 = 81.6 Lbs/Min air × 34.66 ft^3/lb = 2827 CFM at discharge

2827 CFM / .349 ft^2 = 8098 ft/min Max

Min. Vel. 2300 ft/min

Max. Vel. 8098 ft/min

Avg. Vel. 5199 ft/min

Inertia = $MV^2/2g$ = M $(5199)^2$ / 231840 =	117 M
Vertical Lift = MH =	100 M
Horizontal Work = M D f = M 350 × .7 =	245 M
Fittings = MV^2 × π × f N /2g =	1282 M
Total M	1744 M

Air Work = 144($P_1V_1 - P_2V_2$ / Γ- 1) = 144 (26.6963 × 9.84 – 12.6963 × 34.66/1.2-1) = 37467 × 19.74 = 3056392

3056392 lbs/min / 1744 M = 1753 lbs/min or 53 TPH

These are examples of calculating procedures. There are numerous other procedures that may provide different results; the design engineer has to be comfortable with the procedure that is used.

Accordingly, we temper theory with judgment and conservatism. The spread of horsepower ratings available in electric motors gives us ample opportunity to utilize this tempering facility: for example, if the calculations indicate that 12 hp. is required, the use of a 15 hp. Motor provides a 25% safety factor. If we get an answer of 10.5 hp, sufficient conservation exists in the formulas to enable us to specify a 10 hp. Motor without hesitation.

There are other more difficult problems of choice, which can assume considerable proportions, and which justify the existence of the specialist for final design. For example, characteristics of materials can be deceiving. Free-flowing ash, an ideal material to convey, may requires extra power, or an ash with a high calcium content, will require more power and a higher initial conveying velocity, than ash with a lower calcium content.

Then again, things other than purely economic considerations often determine the choice of a system. In some crushing and grinding operations, a negative conveying system will also serve the secondary objective of drawing air through the unit for cooling and cleaning. Chances are that a negative system will win out for such an application, even though a positive pressure system would cost less money.

The pneumatic-conveying specialist can provide still another service in the problem of choice. We haven't said much about disadvantages of pneumatics, but there are a few. The specialist frequently combines pneumatic and mechanical units retaining the function desired, while still effecting considerable savings.

In contemplating final design and installation of a conveyor, any project engineer should consider, or furnish the supplier of the system, with the following information:

1. Distance traversed. This invokes the lift, horizontal distance, and number of bends.

2. Solids characteristics. Size analysis, bulk density, an degree of hygroscopic of the material to be conveyed.

3. Feed considerations. Will the rate of introduction to the conveying line be uniform, or should the supplier provide for uniform feed rate>

4. Automation. To what degree is it desired? Should switches and diverter valves be manual or automatic?

5. Electrical. What are the current characteristics in the plant?

In closing, it might be well to point out that the diversity of special uses pneumatic conveying systems can be put is unlimited.

CHAPTER 15

Common Pneumatic Conveying Equipment

Elbows

Elbows are an important design consideration of any system. Which type of elbow should be used with the various systems is an ongoing discussion between the academic and practical comities. At the present level of agreement, there may never be an answer, but it is really not important that the question be answered. The main thing is to use an elbow configuration that is compatible with the conveyed material, to minimize wear by erosion and to prevent line slugging. With abrasive materials, either sharp-radius elbows or those with an extremely large radius are used. With $r/d \geq 1.5$, wear can be rapid with almost any type of construction. Elbows with intermediate radii should be avoided. First, they will act as centrifugal separators and cause slugging within the system. Second, they may erode very rapidly. When nonabrasive material is conveyed, considerably more latitude is available due to a slower erosion rate. Here the major problem is one of line slugging or pressure surges and can be controlled by using elbows of radii compatible to system velocities. The term slugging and drifting of material after an elbow are almost synonymous if the elbow is located at the base of a vertical riser. Drifting is common after most horizontal bends and after an elbow located at the top of a vertical riser if the elbow is of intermediate radius. The drift may become a slug when the material is re-entrained in the airstream, creating a pressure surge and possible huge thrust loads within the system. Thus elbow radii and conveying velocities to prevent drifting should be used whenever possible.

A sharp-radius elbow (90^0) absorbs most of the velocity energy by impact and bounces the material right back into the air-stream. It causes virtually no pressure surges or drifting in the system, takes up a minimum of space, and is easily replaced, but its use does result in higher pressure losses. When used with a short-wear section of pipe at the outlet of the elbow, it forms a low-cost replacement item. The wear section, 2 to 3 feet long, needs replacing only once every 3 to 10 times an elbow is replaced. For elbows of less than 90 degrees, the replacement is still required less often and is much cheaper than for long- or intermediate-radius elbows.

Traditionally, elbows used in systems handling abrasive materials have been manufactured by each of the suppliers. However, there are now some foundries supplying elbows and other fittings to match those of the major suppliers. They are cast of hard iron in the range of 400 Brinell hardness or better. Historically, most elbows used for negative pressure system handling abrasive materials have been supplied with removable wear backs. See Figure 72. This allows the replacement of a small part of the elbow without removing it or the anchors associated with it. There now is an increasing tendency for suppliers to use solid elbows in place of those with wear backs, largely because the solid elbows are cheaper to produce. It also eliminates one source of leakage. Elbows used for positive pressure systems are generally of solid construction to eliminate leakage through the wear back gasket due to temperature expansion. Standard elbows for most suppliers are 22 ½, 45, and 90 degree,. In addition to elbows, tees and 45 degree laterals are used. See Figure 73 for this type fitting.

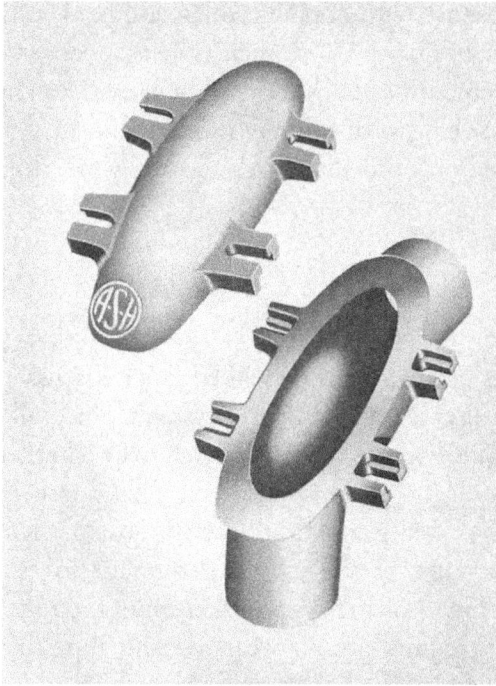

Figure 72

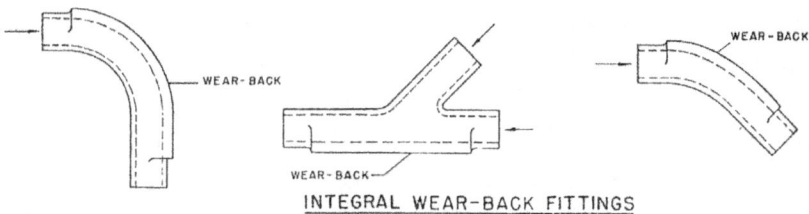

INTEGRAL WEAR-BACK FITTINGS

Figure 73

The use of "Y" type fittings for pneumatic conveying of abrasive materials is discouraged due to high maintenance cost. It is much better, wear-wise, to use a 45 degree lateral and a 45 degree elbow in place of the Y fittings.

For nonabrasive-type applications and for systems where material degradation cannot be tolerated, many special and different elbow configurations are used. A long radius is normally used to prevent degradation. Elbows of rectangular cross section with various types of wear backs and some with concrete fillers are used. On some systems, where the material velocity is low, such as dense phase, or where abrasion is nonexistent, standard steel elbows can be used. Designers must decide if they want to use elbows of extremely long radius to eliminate slugging, or sharp-radius elbows and allow for added pressure drop.

Pipe

There always has been and always will be much discussion of the correct type of pipe to use for a pneumatic system. Since all the pneumatic systems that will be discussed are considered to be completely dry, erosion by abrasion will be the main wear factor. This is largely dependent on the type of material being conveyed, particle size, particle shape, particle hardness, temperature of the material, tonnage to be conveyed, and the operating velocity of the system. Certainly the heavier, harder, and sharper materials require greater wear-resistent pipe than that required for powered material. For abrasive material, the trend is to use cast iron pipe of about 250 Brinell with hard iron wear sections and elbows. It is suggested that carbon steel pipe with hard iron elbows and wear sections can be used for any type of powered material, whether abrasive or nonabrasive. Systems handling nonabrasive materials can use standard steel pipe and elbows. Pipelines handling bottom ash, slag, or other highly abrasive materials require pipe that will withstand considerable abrasion.. The use of basalt-lined and ceramic pipe is becoming very popular. There are many new and different abrasion-resistent pipes on the market. The designer should decide which of the many available products is best to use for the specific application. Pipelines of cast iron or other materials using no rigid couplings require special design consideration for expansion, anchoring support, and gaskets. The pipe normally used for this purpose is available in lengths of 16 to 20 ft maximum. That means there is a joint every 16 to 20 ft and each joint is considered flexible. Depending on the type of joint and gasket used. It is impossible to set a standard that would be broad enough to encompass even a small fraction of the joints used. It is therefore suggested that the designer pay particular attention to the joints of each installation. If the pipe is not flanged or the coupling is not rigidly affixed to the pipe, supporting the pipe on one or both sides of the coupling is required. It is good practice to supply all or part of the supports with "U" –bolts as a minimum. This will allow a support to be a guide as well and prevent buckling of the pipeline due to contraction and expansion.

Provision must be made for expansion within the pipeline due to temperature or movement of equipment attached to the pipe. In some cases the pipeline gaskets are capable of taking the entire movement. Most often, slip-type or bellows-type expansion joints are used. It is important that each end of a straight run of pipe be rigidly supported to allow for the correct movement of the expansion joint.

If the expansion is linearly in one direction, only a single expansion joint is required. This would be true for normal temperature movement. If the expansion occurs in a direction not in line with the centerline of the pipeline, two expansion joints are required to take the movement. Care should be taken in selecting a joint for this application. Bellows-type expansion joints are normally adequate, whereas a slip-type might not be.

Fluidizing

Fluidizing, diffusing, or aeration is probably misunderstood more than any other term used in connection with feeding devices to a pneumatic conveying system. The common belief is that if the material is fluidizable, the addition of air in a controlled manner allows the material to follow the laws of a fluid. In some special instances, this is true, but for the most part, it is not.

The primary use of fluidizing is to promote gravity flow of powered material from a hopper, silo, a day tank, or through some type of feeding devices. In principle, low-pressure air is allowed to permeate the material over a wide area. When this is done, the void area of the material directly over the fluidizers is filled with air at a slight pressure, the angle of repose of material on material is reduced, and material will freely flow by gravity.

When fly ash, lime and most other powered, materials are fluidized in this manner, the angle of repose changes from 90° to about ½ deg.

When air is supplied to a chamber under a porous membrane at ¼ to 15 psi, depending on the height of the material, it is generally sufficient to provide adequate fluidizing. The membrane for best results is of a porous stone that is compatible with the material and has a permeability rating of about 10 cfm per square foot or less at 2" H_2O in air. When fluidizing troughs are set at a slight incline and the material over the trough is fluidized, it will flow by gravity from end to end of the trough. The rate of flow is dependent on the slope of the trough. Tests indicate that flow increases with increases with increasing trough slope up to a maximum rate and remains fairly constant with greater slopes.

Capacity does not vary in direct proportion to the width of a fluidizing trough, but seems to be more of an exponential relation. It is different for each material. Similarly, the air requirements, both in cfm and psi are different on the membrane used and on the material. If the material contains moisture it may be necessary to heat the fluidizing air to prevent localized precipitation of the moisture, causing balling or stickiness at the membrane surface.

When a fluidizing membrane is covered with most powered material it is generally assumed that it will pass air at 5 cfm per square foot of membrane area. This is not an exact figure and is of little importance other than for adequate sizing of the blower for the air supply.

Fluidizers are used primarily to promote flow from hoppers and silos. Silos will be discussed in Chapter 16. See Figure 74 for a typical hopper fluidizer arrangement.

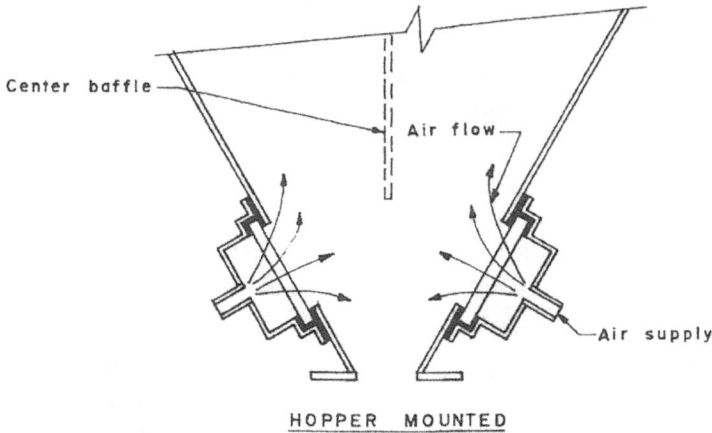

Figure 74

The fluidizing units are located on the sides of the hopper near the outlet. In theory, by promoting flow from the bottom of the hopper, the cavity formed will be large enough that it cannot support the material above and full flow will result. This type of arrangement is most often used with good results. Pressure at 3 psi and air flow at 5 cfm per square foot is normal. In this case, 3 psi is supplied but should be reduced to a minimum that will promote flow. Each material requires a different amount of air and pressures to fluidize it and not all powered materials will fluidize effectively. The designer should test the material to determine the required air inputs. An inert gas can be used just as effectively when the use of air is restricted.

The 3 psi listed above is blower or supply pressure with only ¼ to ½ psi required to promote the flow of most materials. When the air pressure is great enough to cause air to flow through the full head of the material, the material will then act as a fluid, seeking its own level.

Hopper fluidizers are generally used with material temperatures of 350°F or less. Material with temperatures above 600°F is considered to flow readily and not require fluidizing. Between these two temperatures, a gray area exists, with most engineers limiting their use to 350°F. The fluidizing air should be heated to 350°F with the full wall being insulated. This heat is usually supplied by calrod or strip-type heaters, thermostatically controlled to maintain the air space temperature between the wall and the insulation. In all hoppers where material is stored, air circulation is stopped in the lower areas. If there is the slightest moisture content in the material, it may condense on the hopper wall, restricting free flow.

One important concept of fluidizing must be understood. When the height of the material is such that it will seal the flow of air through it, a definite pressure gradient exists, even though there is no apparent action within the material. Once a discharge opening is provided, the material and localized air will flow out of the opening by gravity. The flow will be from the top of the material surface and will take the form of hopper or funnel flow as opposed to mass flow.

CHAPTER 16

Silo Arrangement

Silos are used for temporary storage of powered material. Most often they are of cylindrical shape and elevated for discharge to truck or rail car. The size and construction of a silo depends on its use, and space available, and the type discharge. Most silos are flat bottom, which allows for a number of discharge points such as several rotary unloaders or chutes. This is required when the discharge rate is great enough that it cannot be met with a single point of discharge. Although most silos are constructed of welded steel, concrete, concrete stave, and tile silos are marketed. For materials that must be fluidized, the flat bottom offers a better arrangement; there are some with a conical bottom. In the early years, silos were constructed of vitrified glazed tile, and were limited in size. See Figure 75. These silos were in the range of 20 feet in diameter, with a height of 40 foot of storage. Current designs are as big as 50 to 55 foot diameter, and 100 to 120 feet in height of storage. See Figure 76. The silo is used for temporary storage of ash, and is designed for 2 to 3 days of storage over long week-ends. Depending on the conveying system, vacuum systems will have the collection equipment mounted on top of the silo, with a bag filter to vent the incoming air to atmosphere. For pressure systems there will also be a bag filter for the same purpose, to vent the conveying air to atmosphere.

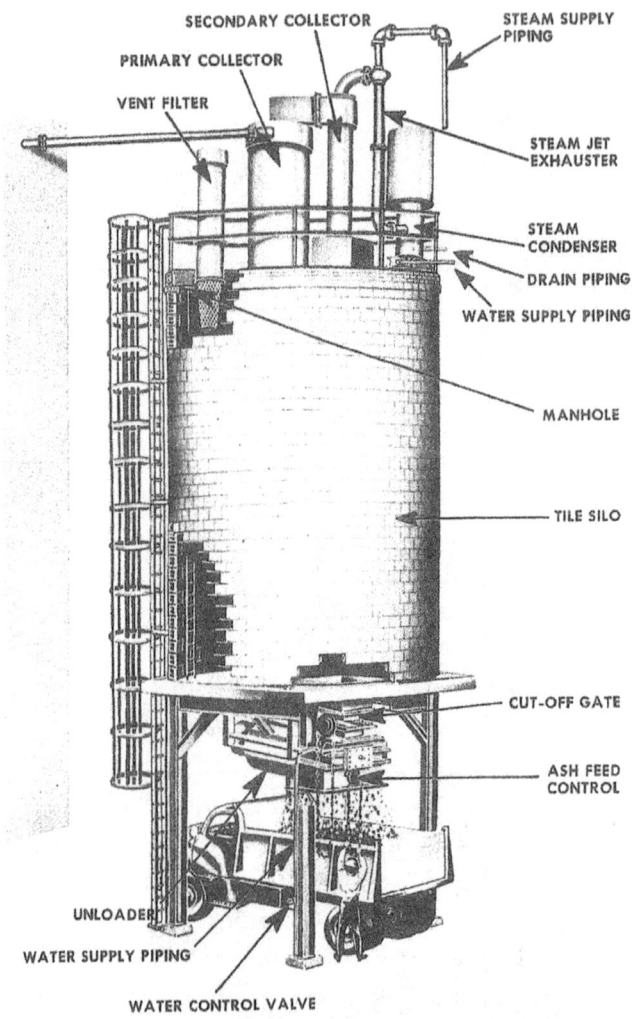

SECONDARY COLLECTOR

PRIMARY COLLECTOR

VENT FILTER

STEAM SUPPLY PIPING

STEAM JET EXHAUSTER

STEAM CONDENSER

DRAIN PIPING

WATER SUPPLY PIPING

MANHOLE

TILE SILO

CUT-OFF GATE

ASH FEED CONTROL

UNLOADER

WATER SUPPLY PIPING

WATER CONTROL VALVE

Figure 75

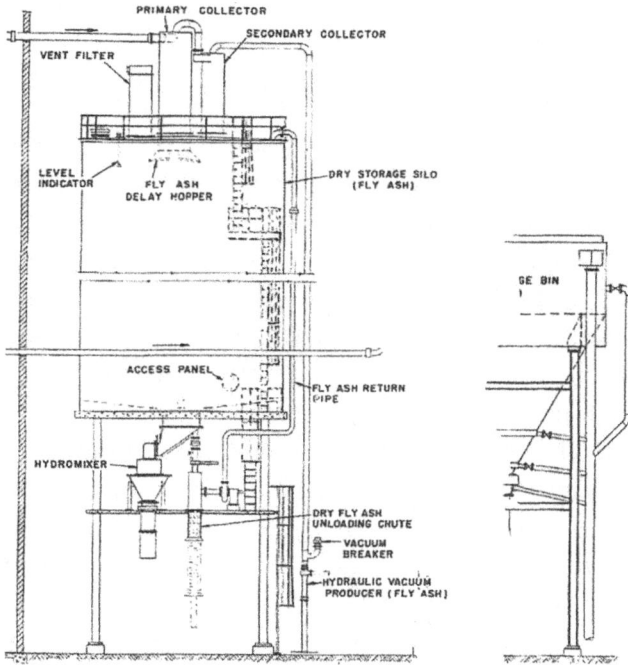

Figure 76

When used with a negative pressure system, dead volume should be allowed for at the top of the silo. Since no air from the conveying system actually enters the silo, the material level may be figured to the maximum limit. As the material drops by gravity from the collecting and separating equipment, it will form a top slope very close to that of the angle of repose of the material. This angle will differ for each material and each height dropped but it is normally set at 45° from the horizontal for all materials.

Allowance must be made for dead storage above the floor of flat bottom silos. If the silo is emptied by gravity, the amount of material that may be discharged is limited by the shape of the cone whose surface by the angle of repose of the material.

The floor of the silo will have a number of air slides to fluidize the ash for unloading. The number of air slides depends on the type ash being stored. The higher calcium ash requires more floor coverage than a lower calcium ash, which means more fluidizing air, and depending on the silo height, more air pressure.

One important concept of fluidizing must be understood. When the height of the material is such that it will seal the flow of air through it, a definite pressure gradient exists, even though there is no apparent action within the material. Once a discharge opening is provided, the material and localized air will flow out of the opening by gravity. The flow will be from the top of the material surface and will take the form of hopper or funnel flow as opposed to mass flow. If a person were to view the top of the material surface when discharge gates were opened, it would be possible to tell how many discharge gates were opened and where they were located. The typical hopper flow discharge pattern starts directly over the discharge point. See Figure 77. When a silo is emptied to the point that full gravity flow is no longer possible, or when the hopper flow effect stems from the discharge opening, further flow from the silo is erratic. Erratic flow cannot be tolerated in many discharge devices so the silo is considered emptied at that point. The designer may wish to set the angle of discharge cone at a value other than 45° to more nearly match the material to be stored and reduce the size silo required.

When positive pressure systems are used for conveying fine-powdered material, the top surface of material in the silo is considered flat because it is fluidized as it is discharged from the conveying system. Conveying air from a positive pressure system does enter the silo; thus a clearance of 5 feet above the material level is normally allowed.

The discharge from a silo is either through a wetting-type unloader, pugmill, or through a dry unloading spout to a closed truck or rail car. The dry unloading spout is used when there is a market for the dry material, such as the use of fly ash in the manufacture of cement.

In a flat bottom silo the compaction forces are basically perpendicular to the silo walls, giving almost no vertical components. Therefore, arching and ratholing are confined to small diameters. With the correct use of fluidizers, cavities are produced around the discharge opening which are too large to support an arch and, as a result, continuous gravity flow is maintained. Generally, the troughs are extended to the outer wall and located so that the floor of the silo is well covered.

Too many fluidizing troughs in the bottom of a silo may actually hinder and retard the rate of discharge. Most of the fluidizing air supplied to the troughs will be discharged with the material. The amount of air that passes through the discharge opening limits the amount of material that can be discharged at the same time. If a rotary unloader is used to wet and condition the material, it can be done more effectively with a smaller dust cloud.

A good knowledge of how fluidizing troughs work will help in laying out a grid for the bottom of any silo. A continuous fluidizing trough is in effect an open: air gravity conveyor and will act in the same manner. It is important that the air gravity conveyors overlap the openings to provide sufficient flow out of the silo. See Figure 78.

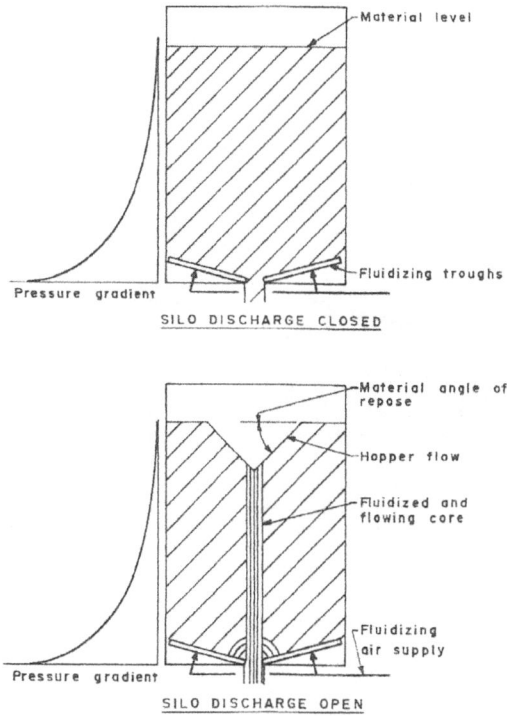

Figure 77

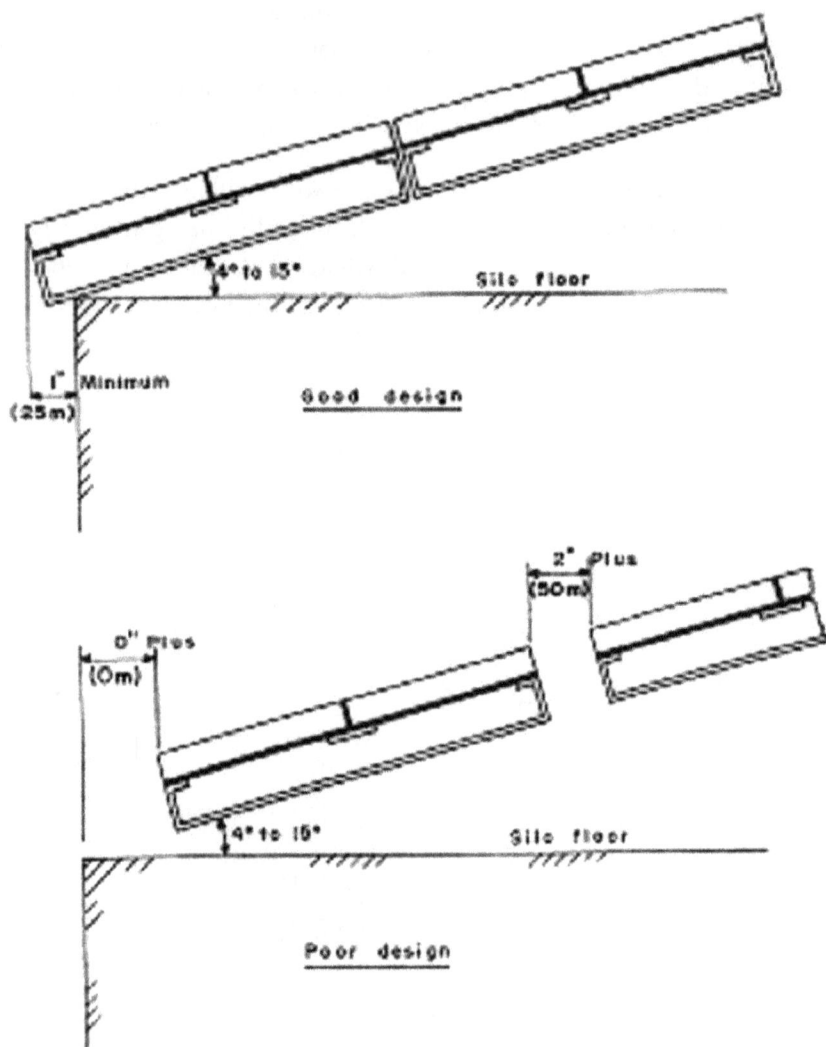

4° to 15°   Silo floor

1" Minimum
(.25m)

Good design

2" Plus
(.50m)

0" Plus
(0m)

4° to 15°   Silo floor

Poor design

Figure 78

CHAPTER 17

Air Gravity Conveyors

One of the most effective and simple conveyors for transporting fluidizable material is the air gravity conveyor. It uses the principles of fluidization and gravity to provide the motive power and has no moving parts. See Figure 79.

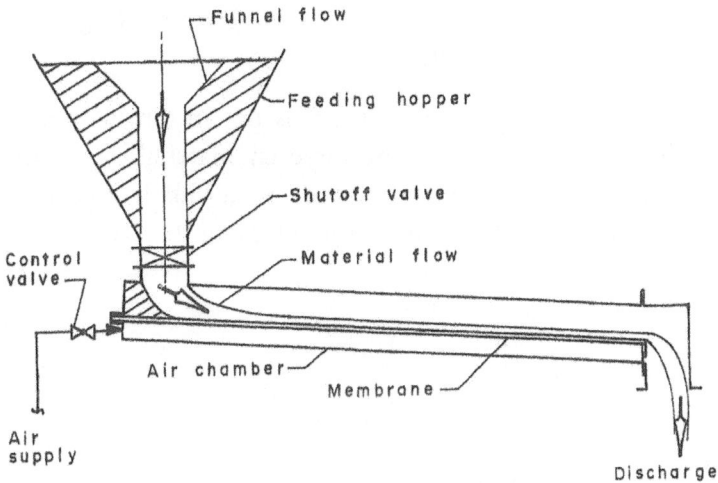

Figure 79

As simple and effective as it is, it is one of the least used conveyors on large scale installations. First it requires a downward slope of about 4 to 10^0 to promote flow, and second, it is adversely affected by damp material. Many conveying applications involve a rise in elevation from the feeding point to the discharge point and it is possible to use an air lift in conjunction with the air slide to provide the rise. This type of conveyor is used more frequently for process control than for any other use.

The familiar aeration pads and fluidizing troughs used on flat or cone bottom silos are, in effect, air gravity conveyors. An air gravity conveyor could be described as a fluidizing trough with or without a boxlike cover from a feeding point to a point of discharge and used to transport fluidizable material by the action of air and gravity.

The aeration membrane is the single most important element of the conveyor. It may be most any material that has a permability rating of about 10 scfm or less per square foot in air at a pressure of 2 inches H_2O, fine grained, and is compatible to the material to be conveyed. This may be in the form of cotton belting, or carborundum stones. The membrane must be heavy enough and strong enough to withstand the constant abrasion and flexing if a fabric is used. It is desirable to operate the conveyor on as little air as possible, to minimize dusting and venting problems at the discharge point.

The capacity of an air gravity conveyor is not in direct proportion to the trough width and is different for each material. Capacity varies exponentially to the trough width. See Figure 80. It increases as the trough slope increases up to a maximum of about 9 or 10°.

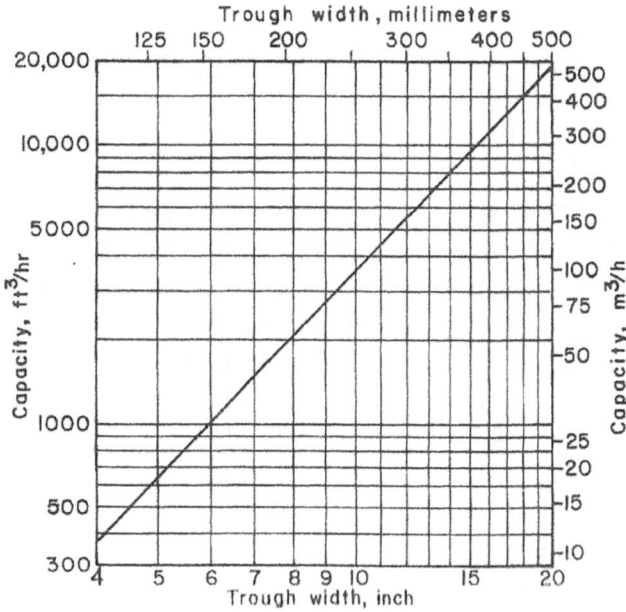

Figure 80

It is fairly constant at greater slopes up to and approaching the angle of repose of the material on the membrane surface. See Figure 81. Capacity of a covered air gravity conveyor is also somewhat dependent on the size of the flow chamber and the amount of air used. If the chamber is not well vented, the capacity of the conveyor will be reduced to the volume occupied by the air.

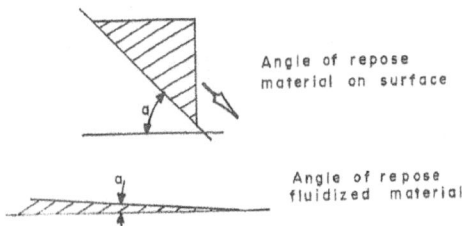

Figure 81

Most any powered material that can be fluidized can be conveyed by an air conveyor. There are many good air gravity conveyors on the market and unless the designer has use for a considerable amount of them, it is probably better to use those of standard manufacture. The membrane should be compatible with the material, including damp material.

One of the drawbacks to the use of gravity conveyors on electrostatic precipitators has been moisture problems associated with boiler startup. The liquid comes from the condensing of the flue gases on the colder plate surfaces of the precipitator hopper walls. With the air gravity conveyor connected directly to the bottom flange of the hopper, the free liquid collecting in the bottom of the hopper causes a plug on the membrane surface. This requires dismantling, cleaning, and at times, even replacement of the membrane. This problem can be partially to totally eliminated by connecting the conveyor to the hopper at a point above the lower flange and channeling the liquid flow away from it. In addition, full air flow should be supplied to the membrane during initial startup and until the precipitator temperature has stabilized and the moisture problem is no longer present. See Figure 82. The water collected may be removed through the flange by removing the flange or at times simply by a drain valve.

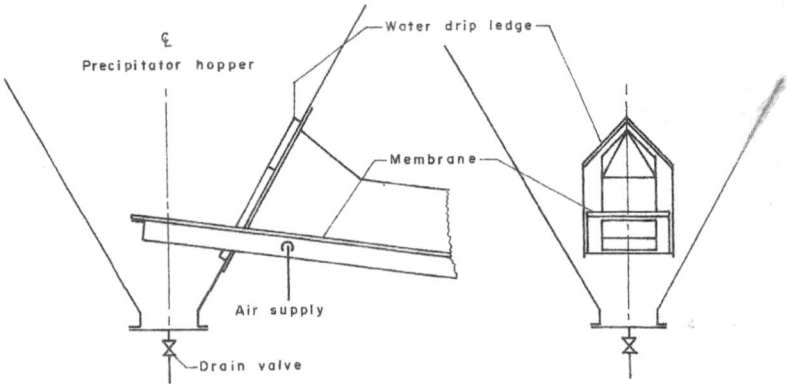

Figure 82

There is no restriction on an air gravity conveyor's length or its configuration. As long as the membrane is continues and the trough has sufficient slope, it may bend or turn any number of degrees without materially effecting its operation. Venting air off the top of the flow chamber or cover can be a problem. Theoretically, the vent connection can be from any spot along the top, but is generally located close to the discharge end. Adequate space must be allowed for the air volume or it will restrict the material flow. The vented air must pass through a filter or to an area that is compatible with the dust.

The air supply should be dry, oil free, and at a temperature that will not cause condensation of moisture in the material. This will often require heating of the fluidizing air to a temperature above the dew point associated with the material. Although an air flow of 5scfm per square foot of membrane surface area at a pressure of 3 psi when covered with material is normally specified, the pressure is usually restricted to a minimum that will promote flow at the conveying rate desired. The required air flow will be different for each material. Air supply of any controllable source may be used. A positive displacement blower, a constant pressure blower, or a central station compressor is acceptable as long as the air is dry.

A thorough understanding of how an air gravity conveyor works and of the controlling factors will help the designer select the correct configuration for any particular use. First, it is a fluidized bed of material throughout its length. Second, the height of the positive head does not necessarily determine or influence the rate of material feed through the conveyor. The conveying rate is influenced by and can be controlled by the height of the inlet opening to the conveyor and by the amount of air supplied to it.

If the material in the feeding hopper is fully fluidized, the material flow chamber will be virtually full of flowing material and air. It will act much like water. This does not say that a high conveying rate can be achieved, as this depends on the amount of air and the restriction to the flow at the discharge of the conveyor.

The correct amount of air is essential to the operation of an Air gravity conveyor. The use of more air than is required will not increase the material flow rate. It will only add to the dust control problem at the discharge end or at the vent connections of the conveyor. It is therefore better to use a minimum amount of controlled air flow to minimize the dust control problems. The volume of air must be considered in designing and sizing the material flow chamber. For long conveyors this chamber should increase in size throughout its length. See Figure 83.

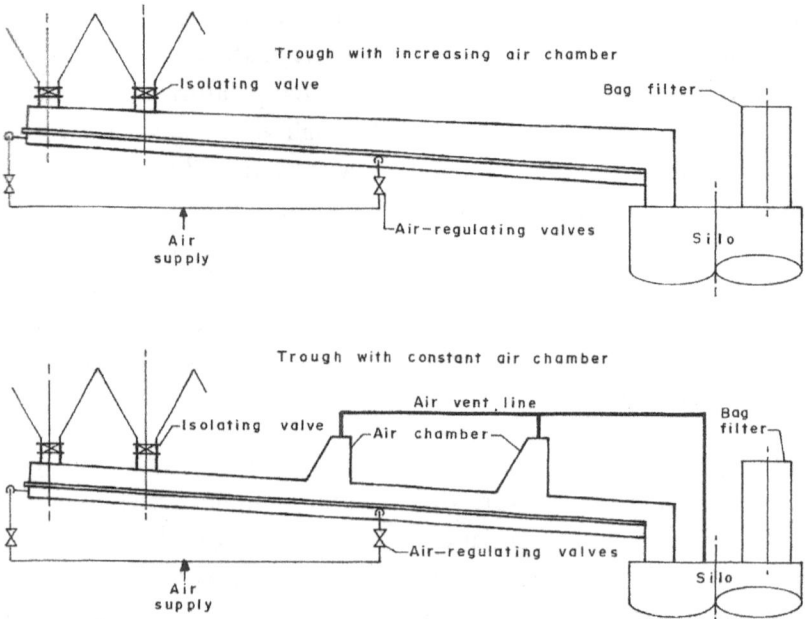

Figure 83

CHAPTER 18

Material Disposal

WET DISPOSAL AREAS AND SLUICE PONDS

For years it was common practice to dispose of ashes from a sluice system into any natural or artificial pond. If the ground is impervious, man-made ponds or lagoons can be constructed in such a manner that water can be recovered, stored and returned to the plant for reuse in the ash handling system. To be effective these storage facilities must cover a considerable area since retention time is the only means by which ash can be settled and separated from the conveying water. Where space permits, volume in a storage basin should be provided for one day's retention of the ash conveying water. Fly ash should not be pumped into a pond from which water is to be reclaimed because of its extremely slow settling rate, unless the pond is greatly enlarged to provide long retention time. There are many human-made disposal ponds in existence whose prime functions are to collect and store the material for a definite period of years, clarify the water, and return it to the system for reuse. They are engineered to perform a definite function. Many of the existing ponds fully meet the existing pollution control laws, although most of them were not designed for that purpose. A clay liner was used primarily to minimize ground seepage so that makeup water to the basic system could be kept to a low value. For a disposal pond to be successful it must be large enough to store the waste product for a given number of years, and it must be adequate to clarify the water so that it can be reused or over-boarded. Even in a climate where there is high evaporation, the pond will eventually overflow due to the displaced volume of material. The overflow must be returned to a natural waterway in a state prescribed by law.

The pollution control laws now in effect detail the quantity and quality of the water that can be returned to a waterway. There is some discrepancy among the federal, state, and local laws, and they are all continually being changed and upgraded. The amount of leachment, into the ground is very limited and in some areas is not permitted at all, especially if the water is in any way toxic. With the present laws, it is virtually impossible to use an untreated site for a new disposal area.

To be acceptable, the disposal pond or ponds for a large industrial complex or an electric utility coal-burning power plant is a complex study and engineering venture. The conditions of no leachment into the ground, and an overflow meeting rigid standards before it can be returned to a natural waterway have caused considerable discussion and expense. The ponds must be lined with either by clay compaction or by the use of impervious liners such as plastic or rubber. Some of the smaller ponds use concrete or asphalt as liners.

For the most part, a disposal pond is built by enclosing an area by dikes. The size is determined by the anticipated production over a period of years, usually the life of the plant. The type of material to be disposed of, and the various materials that will be in the same pond should be known. When there is a large quantity of powered material discharged to the pond, the use of a secondary pond should be considered.

If water is not reclaimed, ponds can be considerably smaller than otherwise. Where ponds are used for temporary storage of material, they should be divided into two compartments so that one side may be cleaned while the other is receiving material. See Figure 84.

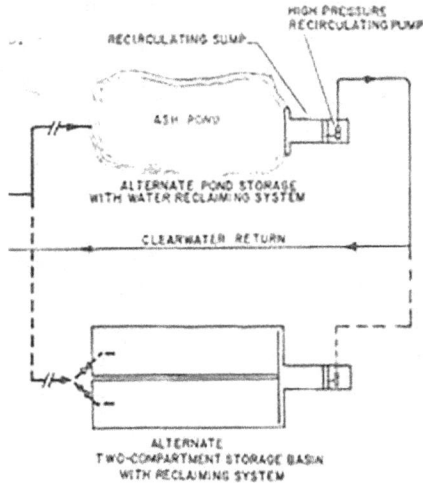

Figure 84

On plant sites where there is insufficient space to allow the use of sluice ponds, smaller dredgible ponds can be used. This is generally accomplished by using two smaller ponds adjacent to each other. They are sized so there is ample time to dewater and completely empty one pond before the second one is full. Dewatering is usually accomplished by pumping the water from the full pond into the empty pond. Most often the material is removed from the ponds by dredging. The material must then be transferred to an above ground landfill.

The designer should consult a text dealing specifically with disposal ponds.

DRY MATERIAL - LANDFILL

The term dry material disposal covers the disposal of any and all material that must be disposed of on a daily basis. It is generally considered to be powered, granular, or chunk material from the boiler plant process. It does not include solid or inert material unless covered by the pollution control laws. These materials can be disposed of in any dump or landfill without regard to air pollution or leachment into a groundwater supply. Every coal burning power plant has a disposal problem for fly ash and bottom ash or slag. For the smaller units, the area needed is far less than the larger units of 500 MW and larger, that can produce around 700 tons of ash per day. Sites to accept quantities of this size should be planned for years in advance.

Until a few years ago, there still aren't many regulating laws covering the disposal of some of the toxic materials. Even with several federal laws and many state and local laws actually conflicting with federal laws, enforcement is not uniformly applied.

The selection and development of a site for these materials is much more involved that it was just a few years ago. A ravine or old quarry is no longer universally acceptable as it was in the past. The first condition that must be addressed is that of ground-water contamination. This means that contamination from the material as disposed of or created by rain or other factors. The site selected must therefore be impervious to leaching or an impervious lining or barrier sufficiently large to contain all leachment must be installed. The second condition is that the disposal of the material after disposal cannot pollute the atmosphere.

These two conditions for the disposal of all waste have made a drastic change in most methods previously used. Any dry landfill for material covered under the laws must meet the minimum standard of, and be approved by, the appropriate bodies before construction is started. For a small fill area this does not present much of a problem. The base of the area and the inside surface of the dikes can be compacted or lined to prevent leachment. The outside of the dikes can be covered with soil and seeded for stabilization. The material can be wetted and compacted at the time of dumping to eliminate dust.

For a large disposal area considerably more work is involved in the initial construction and in its operation. Its location can be any type of terrain that is capable of satisfying the pollution control laws. Most often they are built on flat land and as additional layers are added, their surface is leveled. A well designed landfill will also include a runoff pond for retaining rainwater. It will also have access roads for maintenance.

The first consideration should be the size of land required to dispose of the material for a definite number of years use. The site should be relatively close to the plant to minimize the number of trucks required. Minimum road grades should be used whenever possible so that trucks of smaller size can be used. This will also affect the turnaround time and the total number of trucks required.

Depending on the type of material to be disposed of, its compaction qualities may allow the restraining dikes to be built of it, but only if after compaction it can still meet the pollution laws.

Where ground conditions permit, ashes can be pumped into natural or artificial ponds. If the ground is impervious, man-made ponds or lagoons can be constructed in such a manner that water can be recovered, stored and returned to the plant for reuse in the ash conveying systems. To be effective these storage facilities must cover a considerable area since retention time is the only means by which ash can be settled and separated from the conveying water. Where space permits, volume in a storage basin should be provided for one day's retention of the ash conveying water. Fly ash should not be pumped into a pond from which water is to be reclaimed because of its extremely slow setting rate, unless the pond is greatly enlarged to provide long retention time.

If water is not to be reclaimed, ponds can be considerably smaller than otherwise. Where ponds or pits are used for temporary storage of ash, they should be divided into two compartments so that one side may be cleaned while the other is receiving ash. Overflow from such ponds or pits will flow to sewers or to clarifiers depending on regulations. In most localities untreated overflow is no longer permitted. See Figure 84.

CHAPTER 19

Water Balance

A water balance for any sluice system, regardless of size, may be defined as the total water used per unit time, over an operating period. It is generally set be the supplier of the sluice system and based on the needs of the individual components. However, the designer may estimate the total usage with a good deal of accuracy. It must include the water flow used in the feeding hopper, any impute or outflow at the discharge point, as well as that consumed in the sluice line itself. It will be impossible to cover adequately the many types of conveying systems that could be used in a power plant. As a guide to establishing a working model, a water balance of a jet pump-powered sluice system of a coal-burning boiler will be illustrated. The system will be as shown in Figure 85.

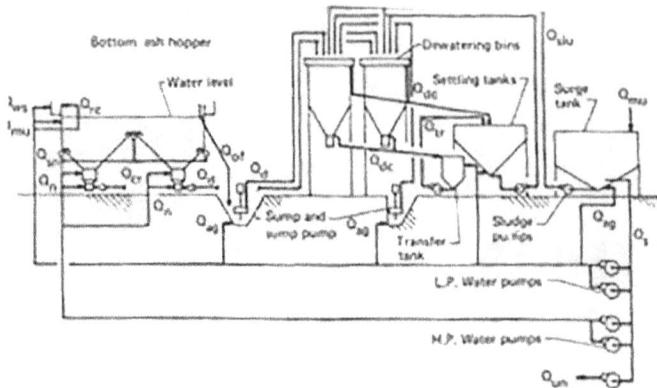

Figure 85

The conveying system will include a water-impounded ash hopper, a jet pump sluice system, dewatering bins, and a complete water recirculation system. The recirculation system to be composed of a setting tank, a serge tank, and all the required pumps and accessories needed for operation. The sluice will operate once every 8 hour and operation of the bottom ash hopper will include the lowering of the water level.

Boiler data:

Powdered coal fired

500 MW capacity

7000 tons of coal per day

10% ash content in the coal

80% fly ash: includes economizer, air heater and precipitator

20% bottom ash: includes pyrites

Sluice system:

12 inch diameter sluice line

8 ft/second sluice line velocity: see table 1, chapter 9

100 tons / hour conveying capacity – bottom ash

Three-compartment bottom ash hopper: 70 ft in length by 11 ft in width

Ash production:

Bottom ash = 7000 (0.10)(0.20 + 0.10) = 210 tons per day

It is common practice in the design and sizing of a bottom ash hopper to add 10% to the specified ash quantity because power plants are built for a life span of 35 years or more. A change in the characteristics of ash produced from coal even from one mine can vary that much over the life span. The individual characteristics of a boiler may produce more or less than the specified percentage of bottom ash.

Net ash volume per shift:

$$V_{net} = (210)\ 2000/3(45) = 3111\ \text{feet}^3$$

This is based on emptying the ash hopper three times per day at an assumed average density of 45 lb/ft^3. Whenever the actual density is know, it should be used. Like everything else density is changeable over time, but it is one of the more stable and obtainable variables. It is generally agreed that the added 10% in ash content will cover any density change.

Gross ash volume per shift:

$$V_{gr} = V_{net}\ (X/8) = 3111\ (12/8) = 4667\ \text{ft}^3$$

It is customary to design the bottom ash hopper on 12 to 16 hour storage so that ample storage time is available should any form of a system stoppage occur (X =12 to 16). This, guards against a forced boiler outage caused by a malfunction to or in the bottom ash hopper of the sluice system, or to any of the inputs to these units. There is also an additional safety factor of 18 in. of cooling water depth above the gross ash volume level. Its purpose is to cool and shatter the ash as it drops into the water.

Without it ash would have a tendency to crust over and form large clinkers. This in itself could cause a boiler outage.

Total volume of a bottom ash hopper:

$$V_{tot} = V_{gr} + (L)(W)(X) = ft^3$$

$$= 4667 + (70)(11)(1.5) = 5822 \ ft^3$$

L = length of ash hopper, ft = 70ft

W = width of ash hopper, ft = 11 ft

X = depth of water over cricket or above the gross ash volume level in feet; this allows for the cooling and shattering of the ash as well as for uneven distribution of the ash in the hopper, X = 1.5 ft

The following typical water quantities and symbols are used for a bottom ash sluice system.

Q_{qs} Water seal, gpm,	30 gpm	
Q_{ec} Refractory cooling,	2 gpm/ft of hopper wall 2(70 + 11) 2 = 324 gpm	
Q_{sn} Slope nozzles, gpm	150 (8) = 1200 gpm eight nozzles at 150 gpm each	
Q_{cr} Clinker grinder seals, gpm	20 gpm/grinder	
Q_d Discharge, gpm (design)	12 in. diameter = 2800 gpm	
Q_n Nozzle, gpm	65 to 85% of Q_d 0.70 (2800) = 1960 gpm	
Q_{of} Overflow to sump, gpm	Q_{ws} + Q_{rc} 50 + 324 = 374 gpm	
Q_{sl} Sludge removal, gpm	400 gpm per system	
	4 in. diameter line at 10 ft/sec	
Q_{ag} Agitation, gpm	150 gpm, settling tank	
	150 gpm, surge tank	

	100 gpm, each sumps
Q_{mu} Makeup, gpm	as required
Q_{un} gpm lost to fly ash	As required
Unloader	
Q_a Average gpm between	300 gpm assumed
new and worn jet pump	
and the system curves	
Q_{dc} Decanting rate, gpm	0 to 3500 gpm
Q_{tr} Rate of transfer	400 gpm
Q_s Suction gpm to the jet	$Q_d + (Q_a/2) = Q_n$
Pump	$2800 + (300/2) - 1960 = 990$ gpm

Time required emptying the first hopper section. When the material is to be removed from the first hopper section of the bottom as hopper, it is necessary first to, remove the supernatant water from the entire hopper. This is the 18 in. depth of water allowed for quenching the ash. It's lower elevation, is set by, and controlled by, the cricket height. As each section is emptied, it is necessary to remove the volume of water equal to the difference between the gross and net volumes of ash plus the volume of water equal to the void volume of the ash itself.

Total water $\quad$ = supernatant + $(V_{gr} - V_{net})$ + voids
$\qquad$ = 7.5 [(70) (11) (1.5) + (4667 − 3111) + (0.5) 3111/3]
$\qquad$ = 16,442 gal

Time to empty water $\quad$ = 16,442/ $(Q_s - Q_{ec})$ = 16,442/(990 − 324)
$\qquad$ = 24.69 min.

Time to empty material = $\quad$ 3111) ft³ (45 lb/ft³) (60 min/hr/
$\qquad$ (2000 lb/ton) (100 ton/hr)3
$\qquad$ = 3111(45) (50/2000)/(100) 3 = 13.99 min

Total time to empty first section − 24.68 + 13.99 = 38.68 min.

In actual operation there is no separation of water flow and material flow. As the hopper discharge door is opened, both material and water are discharged by gravity as a mass. There is no hopper discharge device on the market that will regulate the flow of material, and as a result, the actual rate of discharge of both water and material will be regulated by the jet pump. A jet pump is self-regulating; that is, it has the ability to reject material through the suction port to maintain its discharge pressure and flow. It is for this reason that the nozzle water (Q_n) is set at a minimum of 65 to 85% of the design gpm (Q_d). The percentage is adequate to maintain flow without plugging the sluice line. Also, there will be times when the material flow is less than 100 tons/hr and the system curve will approach that of water. With all these flow variations, it is impossible to arrive at an exact unloading time. Experience has shown that the method of calculation discussed above comes close to actual field test.

Time to empty second and third sections, water only:

Time = $2[(4667 - 3111)/3 + 3111)0.5)/3]$ $(7.48)[990 - 324(2/3)]$ = 20.04 min.

Total time to empty bottom ash hopper:

Time = $24.69 + 20.04 + 3(13.99) = 86.9$ min.

As soon as each hopper section has been emptied and the sluicing operation shifted to line flushing, that section is refilled with makeup water from a low-pressure source.

Hopper refill times are not chargeable to system operating time. The first two sections are refilled with water before the third one is emptied. While the third section is refilled and the supernatant water is returned to the hopper from a low-pressure source, the sluice system continues to convey ash from the pyrite hoppers through a separate holding tank. The amount of low-pressure water used is of importance and as such, the refill time should be calculated for each operation.

Assuming a refill rate of 1500 gpm is used, the total makeup would be equal to (1500 − 323) or 1170 gpm, the 324 gpm being the refractory cooling water rate.

Time to fill one section:

Time = $[3667(7.5)]/[(3)1500]$ = 7.78 min.

Time to refill supernatant water:

Time = $70(11)(1.5)(7.48)/1500$ = 5.76 min.

The example above neglects the effects of water used for refractory flushing and various quantities used for window washing and other miscellaneous values that may be required. Since the total of all these quantities is small, it is considered accurate enough to determine approximate design operating cycles. The actual operating times must be field set. Evaporation from the bottom ash hopper, settling tank and the surge tank is normally neglected, as it will only show up in the system makeup to the surge tank. It is a minor value and results in a slightly longer operating time for the makeup pumps. If the makeup quantity is limited, it should be calculated, especially if the system is located in an area of low humidity.

Maximum high-pressure water usage occurs during the operation of the slope nozzles and is 2560 gpm. It is suggested that the pump be specified with this gpm plus a bypass allowance, or that automatic bypass valves be installed to prevent pump operation at a closed-off head.

FIGURES

1. Overall view of Power lant
2. Shows An Old Style Ash System
3. Shows Ash Being Removed From Ash Hopper into Sluice System
4. Sluice Liners with turnbuckles for installation
5. Sluice Showing multiple trenches
6. Sluice liner showing nozzle in a stepped arrangement
7. Sluice liner showing nozzle in a continuous arrangement
8. Centrifugal Pump
9. Jet Pump
10. Jet Pump Curves
11. Showing Various Centrifugal Pumping Setups
12. Showing Simple Level Controls at Transfer Tank
13. Overview of Mill Reject System and Ash Hopper and Ash Ponds
14. Water Impounded Ash Hopper
15. Hopper With Discharge at End
16. Hopper With Discharge at Middle
17. Slag Tank Arrangement
18. Gate House Arrangement
19. Typical High Capacity Clinker Grinder Arrangement
20. Typical 2 Foot Clinker Grinder

www.ingramcontent.com/pod-product-compliance
Lightning Source LLC
Chambersburg PA
CBHW031856200326
41597CB00012B/434